Soil Science Simplified

Sixth Edition

Soil Science Simplified

Sixth Edition

Neal S. Eash
*Professor of Biosystems Engineering &
Soil Science
University of Tennessee in Knoxville,
Tennessee.*

Thomas J. Sauer
*Research Soil Scientist in the Soil,
Water, and Air Research Unit at the
USDA Agricultural Research Service
in Ames, IA, USA.*

Deb O'Dell
*Doctoral Candidate in the Biosystems
Engineering & Soil Science
Department at the University of
Tennessee in Knoxville, TN, USA.*

Evah Odoi
*Adjunct Assistant Professor
in the Biosystems Engineering & Soil
Science at the University of Tennessee
in Knoxville, TN, USA.*

Illustrated by Mary C. Bratz

WILEY Blackwell

Published by John Wiley & Sons, Inc., Hoboken, New Jersey
Published simultaneously in Canada

For general information on our other products and services or for technical support, please contact
our Customer Care Department within the United States at (800) 762-2974, outside the United States
at (317) 572-3993 or fax (317) 572-4002.

Wiley also publishes its books in a variety of electronic formats. Some content that appears in print
may not be available in electronic formats. For more information about Wiley products, visit our web
site at www.wiley.com.

Library of Congress Cataloging-in-Publication applied for.

Cover image: Mycola, wet spot on cracked earth under dramatic sky, iStock / Getty Images Plus
Typeset in 9/11pt TimesLTStd by SPi Global, Chennai, India

1   2016

CONTENTS

PREFACE

Soil Science Simplified, Sixth Edition explains soil science in an easily understandable manner. Students, professionals, and nonprofessionals alike will gain an accurate working knowledge of the many aspects of soil science and be able to apply the information to their endeavors. The book is a proven and successful textbook and works well as assigned reading for university students in the natural sciences and earth sciences. Agricultural science courses taught at the high school or post high school level can also use this edition as a resource.

Soil science has been largely directed toward agricultural production. Farming remains at the forefront of food and fiber production and is, more than ever, concerned with soil and its properties.

Anyone who works with soil can benefit from an understanding of soil and its properties. Horticulturists, foresters, landscape architects, and similar professionals can benefit from an in-depth understanding of soils. Home gardeners can likewise benefit. Those who construct houses and other structures need to understand that the soil's physical and chemical properties can impact foundation problems. Engineers need the same understanding of soil properties as they build roads, bridges, dams, levees, and similar structures. Environmentalists and people in related areas find a working knowledge of soils useful.

There are many uses of soil-far more than for production agriculture. And everybody who works with the land in any way needs to know how to take full advantage of the information in a soil survey report. The need for an understanding of soil is ever-present. If your profession will involve the use of soil, read and understand the information in this sixth edition of *Soil Science Simplified*. Keep a reference copy in a handy spot in your bookcase.

This sixth edition expands and updates several chapters. New approaches to the content have been incorporated to provide information needed by those professionals listed previously. A chapter on conservation agriculture (CA) has been added that describes the evolution of agricultural management practices that support and strengthen both food production and environmental resources. The illustrations and photos demonstrate the principles described in the text and enhance comprehension.

Drs. Eash, Sauer, and Odoi are experienced university professors of soil science who have taught and conducted research in soils. Through experience in the field, classroom, and laboratory, they have gained a basic, hands-on appreciation of the importance of applied soil science. This book represents their many years of experience and the desire to provide a working knowledge of soil and how its properties influence decisions on the best use of soil, whether it is used as a medium for plant growth, as a base for the foundation of buildings, or for any other purpose.

This book has been used successfully as a resource in certification programs in the agricultural industry such as the Certified Crop Advisor program sponsored by the American Society of Agronomy. We once again use many of the line illustrations by Mary C. Bratz that have appeared in earlier editions of the book which continue to be useful in communicating essential ideas and processes in soil science.

CHAPTER 1

Introduction to Soil

Soil is a natural resource on which people are dependent in many ways. Since the birth of the soil conservation movement in the 1930s, there has been an increased interest in conserving the soil. The environmental awareness and concerns that have occurred over the past several decades have focused attention on the need to conserve soil as a fundamental part of the ecosystem. There is, however, little public understanding of the soil's complexity.

Careful observers may see soil exposed in roadbanks or excavations, and it may be noticed that the soil does not look the same in all locations (Fig. 1.1). Sometimes the differences are apparent in the few inches of surface soil that the farmers plow, but greater variations can usually be seen by looking at a cross section of the top 3 or 4 ft. (0.9 or 1.2 m) of soil. The quality and quantity of vegetative growth depends on the properties of the soil layers.

Roads and structures may fail if they are constructed on soils with undesirable characteristics. Special care must be taken to overcome soil limitations for specific engineering uses. Satisfactory disposal of human waste and livestock manure is becoming an increasing concern, particularly where soils are used as a disposal site.

Poor yields of agricultural crops and poor growth of trees may result from a mismatching of crops and soils. This mismatching may happen because the landowner has not examined the soil horizons or understood their limitations. Soil scientists study the factors necessary for proper soil management and plant growth.

What Is Soil?

The traditional meaning of soil is that it is the natural medium for the growth of land plants. The Soil Science Society of America has published two definitions. One is "The unconsolidated mineral or organic material on the immediate surface of the earth that serves as a natural medium for the growth of land plants."

A more inclusive definition by the Society is "The unconsolidated mineral or organic matter on the surface of the earth that has been subjected to and shows the effects of genetic and environmental factors of: climate (including water and temperature effects) and macro- and microorganisms, conditioned by relief, acting on parent material over a period of time."

Soil Science Simplified, Sixth Edition. Neal S. Eash, Thomas J. Sauer, Deb O'Dell, and Evah Odoi.
© 2016 John Wiley & Sons, Inc. Published 2016 by John Wiley & Sons, Inc.

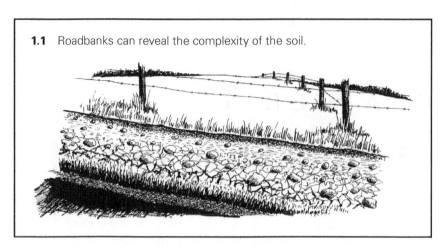

1.1   Roadbanks can reveal the complexity of the soil.

The effect of each of these genetic and environmental factors will be discussed in Chapter 2 on soil formation.

Soil differs from the material from which it is derived in many physical, chemical, biological, and morphological properties and characteristics. The differences in these properties and characteristics will be discussed in subsequent chapters. Their effect on soil management decisions is important whether the soil is to be used for crop production, in an urban setting, or for roads, dams, waste disposal, and its many other uses.

Most soil consists of fragmented and chemically weathered rock which includes sand, silt, and clay separates, and it usually contains humus, which is partially decomposed organic matter. Soil is very diverse over the face of the earth, and it varies considerably. If properties of a soil are known, the soil can be properly managed, and it will serve quite well for the purpose for which it is used.

Nature and Uses of Soil

Soil is a medium in which plants are grown for food and fiber. It is fortunate that over most of the land area of the earth, soil covers bedrock to a considerable depth. If there were no soil, the continents would be wastelands of barren rock. In soil, seeds germinate and plants grow as they obtain water and nutrients from the soil. Crops of the fields and forests produce food and fiber.

Soil gives mechanical support for plant roots so that even tall trees stand for decades against strong winds. Soil also physically supports structures such as houses, buildings, sidewalks, streets, and highways. Sometimes the properties of soils are undesirable and buildings and pavement will crack due to the instability of the underlying soil. Abandoned roads may be buried by soil carried upward by ants, earthworms, and other creatures.

In intertropical regions, millions of people live comfortably in places built chiefly from locally excavated soil. Such earthen houses are common in West Africa (Fig. 1.2). A compound earthen dwelling for an extended family may be quite an impressive structure.

1.2 Earthen houses are common in West Africa.

The adobe houses of the southwestern United States and the pioneer sod houses of the prairies are other examples of earthen houses. Modern earth-sheltered homes are shown with pride by the owners and builders. For maximum insulation, houses often feature an earthen embankment covering all but the side exposed to the sun.

Soil is involved in several processes in the hydrologic cycle. Water in the form of rain, dew, fog, irrigation, or snowmelt may move into the soil (infiltrate) or evaporate or run off of the soil surface into the area drainage system into lakes or streams. The water that infiltrates into the soil may evaporate or be utilized by plants. It may be used by the plants to form compounds or it may be transpired from the leaves back into the atmosphere. If there is more water than the soil will hold, it may percolate downward to become part of the groundwater reservoir and eventually become part of streams, rivers, and springs.

Many organic and inorganic pollutants in wastewater are sorbed as they pass through the soil, thereby partially cleansing the groundwater. If potential pollutants that are very soluble are added to the soil, they may be carried by the soil water into the groundwater to our detriment.

Soil is an air-storage facility. Plant roots and billions of other organisms living in the soil need oxygen. The pore system in soil provides access to air, which moves into and drawn out of the soil by changes in barometric pressure, by turbulent wind, by the flushing action of rainwater, and by diffusion. Some plants, such as rice, have the capacity to conduct oxygen into waterlogged soil. Soil air contains considerable amounts of carbon dioxide.

Soil is even useful as a mineral supplement for people. In some impoverished African countries, selected types of soil containing high calcium, for example, have been used as a special food supplement. Specifically, pregnant women and their babies have benefited from the mother's ingestion of soil from termite mounds that are enriched with calcium. By using this natural resource, these women may have adequate calcium in their systems.

Soil accepts back that which came from it. When plants die, it is not long before organisms that cause them to decompose will be active and the plant will eventually become a part of the soil. Even huge logs on the forest floor soon disappear (Fig. 1.3). Animals that live in the wild as well as other forms of life also return to the soil when they die. Society

1.3   Living organisms sooner or later become a part of the soil once again.

produces vast amounts of waste of every size, shape, and description, which is often buried in landfills where it will decompose if it is organic.

Soil is beautiful; it is an aesthetic resource. People may become fond of their native soil, whether it is black and brown or red and yellow. There is a rainbow of various hues of soil under our feet.

Changes in both soil and vegetation through the seasons are observed with great interest. Some soils form wide cracks in dry seasons and swell when the rains return. Frost action may create little ice pillars that lift the surface of the ground in winter. The smell of freshly tilled soil seems good to farmers and gardeners as they plant their crops with high expectations for an abundant harvest. Some people love their native soil so much that even today they still perform the ancient ritual of kneeling to kiss it when they return home.

How Big Is an Acre? A Hectare?

Land measurements in the United States are in the English system while for most of the rest of the world the metric system is used. The principal measure of land in the English system is the acre while the hectare is used in the metric system. One acre equals 0.4047 ha; hence, an acre is only about 40% the size of an hectare. An acre has 43,560 sq ft, while a hectare contains 107,628 sq ft.

In the United States, soil amendments are usually applied in units of pounds (or tons) per acre. In the rest of the world, applications are usually in metric units of kilograms (or metric tons) per hectare. Pounds per acre closely correspond to the metric units of kilograms per hectare. Tons per acre closely correspond to metric tons per hectare. Loss of soil by erosion is rated in tons per acre.

1.4 An acre is 208.7 ft on a side; a hectare is 328 ft on a side.

Comparing a house lot size to an acre will give a perspective on the size of an acre. Most house lots will be approximately 10,000 sq ft in size (about one-fourth of an acre), whereas an acre has 43,560 sq ft. See Figure 1.4 for a perspective on these units of measurements.

CHAPTER 2

Soil Formation

Many people throughout the world depend upon soil for their subsistence. Soil is a dynamic feature on the landscape and few realize the importance of the parent rock to soil fertility and productivity. Although the parent rocks determine many soil characteristics, soil is much more than just weathered rock.

Pedogenesis is the term used to describe the formation and development of the soil profile. The **"pedon"** is the three-dimensional body of soil used as the soil base of reference and **"genesis"** is often defined as "beginning." Through pedogenic processes the soil profile develops from the very thin—maybe several inches thick which is common in young soils—to soil that is greater than 80 in. (2 m) thick, which is common in older soils.

To understand soil pedogenic processes, it is essential to consider how rocks are formed, how that formation influences their mineralogy, and subsequent breakdown into soil parent material. The minerals in rocks strongly influence the composition of the soil derived from them. The other factors in nature that influence the specific properties of soil in a given location will also be discussed in this chapter.

One of the basic rules of nature is that nothing remains the same over long periods of time. Astronomers tell us that even stars such as our sun have a finite lifespan. They coalesce from cosmic dust, form into shining solar bodies, finally expend their energy, collapse, and return to cosmic dust. The secrets of these processes have only recently been revealed by the Hubble telescope. On earth, the alteration of rocks from one form to another is much more easily understood because we can study specimens of rocks and relate them to their position in the earth's crust.

Rocks are merely combinations of minerals. Minerals have specific chemical compositions whereas a rock refers to a material within a specified range of mineralogical composition that is of appreciable extent in the crust of the earth. Some of the most common rocks are granite, basalt, sandstone, and limestone. Rounded pieces of rock—so common in glaciated regions—are boulders, stones, cobbles, and gravels in descending order of size.

The Rock Cycle

To understand the formation of soil, consider first the rocks from which the mineral particles in the soil were derived. As the earth cooled, the molten magma crystallized into

Soil Science Simplified, Sixth Edition. Neal S. Eash, Thomas J. Sauer, Deb O'Dell, and Evah Odoi.
© 2016 John Wiley & Sons, Inc. Published 2016 by John Wiley & Sons, Inc.

2.1 The rock cycle shows how heat and pressure, melting and erosion cause rocks to change in form through geologic time.

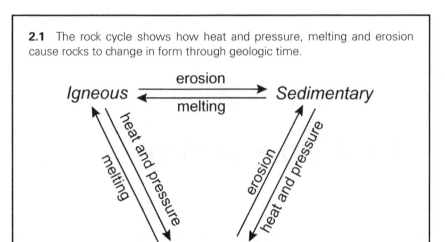

igneous rocks. As long as there has been water on the earth, flowing water has been eroding rocks and the fine particles produced have become sediments, which may solidify into sedimentary rocks. Under conditions of extreme heat and pressure, both igneous and sedimentary rocks may be modified and at least partially recrystallized into metamorphic rocks.

The shifting of continents causes landmasses to slide over and bury other landmasses to the extent that the buried ones may become molten again. Where this occurs there is evidence of great tectonic activity in the form of earthquakes, volcanoes, faults, and related phenomena. Therefore, over geologic time the rocks of the earth are cycled from one form to another (Fig. 2.1). Rocks are the evidence for these actions in the past, and the same processes continue today.

Composition of the Earth's Crust

Chemists recognize a few over 100 elements that make up everything tangible on earth. Of these, the eight listed in Table 2.1 are the most abundant elements in the earth's crust. The others are no less important, but are present in much smaller quantities.

Silicate Minerals

If molten magma from within the earth cools very rapidly, these elements solidify randomly into a glass such as obsidian, a material commonly used in jewelry. If the cooling is slower, the elements will assemble themselves into crystalline silicate minerals. The slower the cooling is, the larger the crystals.

Silicates are minerals made up, in large measure, of combined silicon and oxygen. They are the most common minerals in rocks. When only silicon and oxygen ions are involved, they form a four-sided structure with oxygen ions at the points and a silicon ion in the center. It can be compared to a three-sided pyramid, with the base being the fourth side. This is called a tetrahedron. If the O^{2-} on each corner is shared with another tetrahedron, a very strong framework structure results. A mineral with this form is quartz, and it is so resistant that it is said to be nonweatherable. Hence, the beaches along our oceans consist mainly of sand.

Of the silicates, most are **aluminosilicates;** feldspars are the classic example. Feldspars also have a framework structure but from one-fourth to one-half of the Si^{4+} was replaced with Al^{3+} during the original crystallization of the feldspar. Since Al^{3+} has a lower positive charge than Si^{4+}, the

Table 2.1 Composition of earth's surface crust

Element	Ion
Oxygen	O^{2-}
Silicon	Si^{4+}
Aluminum	Al^{3+}
Iron	Fe^{2+}, Fe^{3+}
Calcium	Ca^{2+}
Magnesium	Mg^{2+}
Potassium	$K+$
Sodium	$Na+$

unsatisfied negative bonds from the O^{2-} are satisfied primarily by K^+ and Ca^{2+} in the crystal. Feldspars are quite stable but are less resistant to weathering than is quartz. The weathering of feldspar accounts for much of the potassium and calcium found in the soil, the oceans, and sedimentary rocks.

Micas are the other main group of aluminosilicates. The tetrahedra are formed into layers that can be lifted, one from the other, like the pages of a book. When separated from the rocks, these small flat particles will glisten in the sun, especially if they settled out of flowing water and lay flat on the dried soil surface.

Most of the very dark colored minerals in rocks are ferromagnesian silicates. Instead of the framework silicate structure discussed above, these minerals have single, paired, or chained sets of tetrahedra that are bonded together by accessory ions, usually Fe^{2+} and Mg^{2+}, hence, the term ferromagnesian. It is by way of the accessory ions that weathering gains access to these minerals and the integrity of the mineral structure is destroyed. Two common groups of these dark minerals are the amphiboles and pyroxenes.

Igneous Rocks

Igneous rocks (Fig. 2.2), including granites and their metamorphic associates, make up the bedrock foundation of the continents. The minerals in them are crystalline

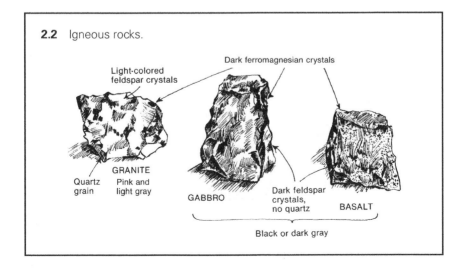

2.2 Igneous rocks.

Light-colored feldspar crystals

Dark ferromagnesian crystals

GRANITE

Quartz grain

Pink and light gray

GABBRO

Dark feldspar crystals, no quartz

BASALT

Black or dark gray

in form and, if the magma cooled slowly far below the surface of the earth, the crystals are comparatively large. This is the case with granite. If the cooling of the magma took place more rapidly, the crystals are small, such as in rhyolite. Granite and rhyolite may be identical in mineralogical composition and are characterized by having abundant quartz due to the high silica content of the magma. In a parallel manner, magma lower in silica may solidify into very dark colored gabbro or basalt, depending on the rate of cooling.

Crystalline igneous rock lays just below the unconsolidated surface material on about one-quarter of the earth's land area. Elsewhere it is more deeply buried. It is quarried for building stones and monuments. Pink and light-colored granite is popular. It outcrops dramatically in the Black Hills of South Dakota at Mount Rushmore, where the heads of four U.S. presidents have been carved. Gabbro can be polished into a beautiful building stone and is sometimes called black granite. A well-known example of it is the Vietnam Memorial in Washington, DC. Blackish and finely crystallized basalt is well known because of extensive volcanic activity on earth.

Sedimentary Rocks

Sedimentary rocks (Fig. 2.3) are the bedrock for about three-quarters of the land area of the earth. These rocks were deposited as loose layers of sediment on the bottoms and edges of ancient seas. Sand, primarily quartz grains, was deposited near the shores, gray siliceous mud farther out, and limy, whitish mud from fossil shells in the deep water. These layers gradually hardened into rock to become sandstone, shale, and limestone, respectively. As the land was slowly uplifted and the seas receded, sedimentary rock covered most of the continents.

Metamorphic Rocks

Rocks can be altered by heat and pressure within the earth. The **metamorphic** rocks that result may have been any of the igneous or sedimentary rocks. Granite is commonly metamorphosed into gneiss, a beautifully banded rock wherein like minerals became concentrated due to similar **viscosity** and **density** in the shifting magma. Sandstone is cemented by silica from solution to become quartzite, which is the most resistant rock that is widespread on the earth. Shale is converted into slate and limestone into marble by heat and pressure.

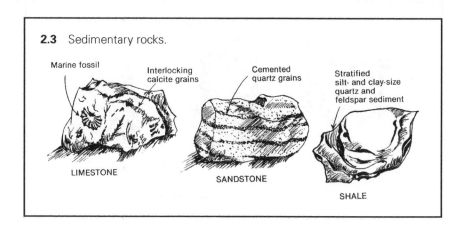

2.3   Sedimentary rocks.

Marine fossil

Interlocking calcite grains

Cemented quartz grains

Stratified silt- and clay-size quartz and feldspar sediment

LIMESTONE

SANDSTONE

SHALE

Processes of Rock Weathering

When living organisms such as plants die, they are rotted by saprophytic microorganisms. In a similar manner, naturally occurring physical and chemical forces cause rocks to be weathered into saprolite. Collectively, saprolite is called the **regolith** of the earth, which is composed of the loose mineral materials above solid bedrock. The effects of rock weathering can be observed by splitting a stone that has been exposed on or near the ground surface for a long time (Fig. 2.4). During the weathering process, the altered rock material may accumulate in place over

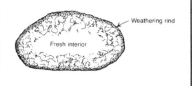

2.4 Exposure to weathering causes tiny cracks to develop in the surface of rocks, which allows for chemical reactions with the penetrating solutions.

Weathering rind

Fresh interior

the solid rock or it may slide, be washed, or be blown to other sites. Soil formation begins soon after loose rock material is stabilized.

Biological Decomposition of Rocks

If you have ever been on a hike into the mountains and sit down on a rock to rest, if you look closely at the rock you will quickly see that it is covered by small macroscopic organisms called **lichens.** Lichens are composed of an alga and a fungus and are often one of the first organisms to colonize exposed rocks.

Just three years after the island of Krakatoa was largely blown away by a violent volcanic eruption in 1883, scientists visited it and found that the surface of the fresh bedrock was already being invaded by cyanobacteria, one of the most self-supporting forms of life on earth. It can both photosynthesize and fix nitrogen. Growing along with the cyanobacteria were nitrogen- and carbon-fixing bacteria as well as fungi and lichens. Weak acids produced by these microorganisms were dissolving nutrients (phosphorus, calcium, etc.) from the rocks and building up a humic mat capable of supporting mosses and eventually higher plants. The weak acids include carbonic acid formed by solution of carbon dioxide gas in water and lactic acid produced by fungi, and the stronger acids (nitric and sulfuric) were formed by bacteria. Certain fungi and bacteria can release phosphorus from mineral particles. It is evident that microorganisms are involved in rock weathering from the start.

Physical Weathering

Physical weathering of rocks is their breakdown into progressively smaller pieces with no change in molecular arrangement within the minerals. Any of the forces that transport solid particles causes them to wear. Sand on a beach that is rolled by each incoming wave is a familiar example. Strong winds pick up sand and blast it against objects that soon show the effects of abrasion. Tree roots penetrate cracks in rocks and as the roots grow they cause the cracks to expand and eventually break the rock. In temperate regions, water enters cracks in rocks, freezes, and can cause the surface of the rock to peel off like the rind of an orange, which is called exfoliation. Glaciers were the ultimate in physical weathering as they broke loose massive boulders and moved them great distances with a grinding action. Hills were lowered, valleys were filled, and there was a general leveling effect except at the glacier's edge, where small hills called terminal moraines were created. No matter the extent of physical weathering, it does not directly cause significant release of ions from the minerals for the benefit of plants.

Chemical Weathering

Chemical weathering, as the term implies, results from chemical reactions that alter the molecular composition of minerals. These chemical forces react with the surface of minerals. If physical weathering did not greatly increase surface area by breaking down rock into smaller pieces, chemical weathering would progress much more slowly.

Hydrolysis. Hydrolysis is important in mineral weathering. It takes place when hydrogen ions (H^+) in water replace metallic ions in minerals. All water is slightly ionized, so hydrolysis is pervasive. Rain absorbs carbon dioxide (CO_2) as it falls, resulting in a relatively weak carbonic acid (H_2CO_3), which greatly increases the reaction of hydrolysis. However, the ancient statues from the Greek and Roman empires did not show much degradation until smoke from modern industry resulted in sulfuric acid and nitric acid in the precipitation. Over millions of years, most of the acid in the soil resulted from the respiration of CO_2 by living organisms. Plant roots also release H^+ during nutrient uptake.

In one simplified example of hydrolysis, potassium feldspar reacts with water to yield silicic acid and potassium hydroxide as shown in following equation:

$$KAlSi_3O_8 + HOH = HAlSi_3O_8 + KOH$$

The silicic acid is the building block of clay. In the reaction, the primary mineral is destroyed, clay is formed, and potassium (K^+) is released into the soil for use by plants. If precipitation is sufficient to leach away the base (KOH), the land will become more acid and the sea will become more basic.

Oxidation. Oxidation takes place when certain multivalent ions lose an electron (a negative charge) to become more positive. A common element in rocks capable of two valence states is iron. Just as a wrench left in the rain will rust, so also iron-bearing minerals in rocks become oxidized. The equations for the reaction are

$$Fe^2+ = Fe^{3+} + 1e^- \text{ and } 4\,Fe^{3+} + 3O_2 = 2Fe_2O_3$$

Hydration. By itself, oxidation would not be extremely disruptive to the mineral, but, in nature, it is followed by hydration:

$$Fe_2O_3 + nH_2O = Fe_2O_3 \bullet nH_2O$$

In this reaction, n water molecules attach themselves to an iron oxide molecule. This results in considerable expansion, which greatly disrupts the mineral structure of the rocks and causes them to crumble. For this reason, when digging in the subsoil in a humid region it is common to encounter stones that disintegrate if struck by a spade. Manganese is another element in minerals that can exist in multivalent ionic states, but it is much less abundant than iron. Salt may also hydrate with similar results.

Reduction. Reduction, being the opposite of oxidation, reflects a gain of electrons in multivalent ions. It is not disruptive to most bedrock, but it does have a marked influence on soil where oxygen has been depleted by microorganisms in wet places. Under reducing conditions, iron and manganese may be dissolved and removed from the system or translocated to regions with free oxygen. Here they precipitate as nodules, concretions, or various types of layers and coatings.

Solution. Water comes as close as anything to a universal solvent. However, it is only capable of dissolving large quantities of soluble salts that were precipitated from solution at some earlier time. The calcium carbonate in limestone came from the shells of sea creatures. It is an example of a salt that is slowly soluble in pure water, but water enriched by carbonic acid, due to biological activity, reacts with limestone. This dissolution is evidenced by massive caves where the rock was dissolved by biologically acidified water that seeped down from the soil.

Factors of Soil Formation

Soil scientists think of soils as natural bodies that have length, breadth, and depth. Each soil body occupies a portion of the landscape. This means that soils are more than simply the product of rock weathering; they are components of the landscape (Fig. 2.5), just as are rivers, forests, marshes, and prairies. Thousands of years have been required to make our present-day soils. Five factors of soil formation have been identified (Fig. 2.6). They are (1) parent material, (2) climate, (3) living organisms, (4) topography, and (5) time.

Soil Parent Material

Parent material of mineral soils is the weathered rock that was slowly broken up at a site or was transported there by natural agents. It can be grouped into (1) **crystalline** rocks, such as granite and gneiss, (2) **sedimentary** rocks, such as sandstone and limestone, and (3) geologically recent **deposits,** such as alluvium and glacial till.

Soils that have formed from granite contain a full range of particle sizes, from gravel and sand to the finest clay. Since quartz grains (somewhat like bits of glass) in granite are very resistant to weathering, they become the gritty sand in the soil. The less-resistant minerals in rock—such as feldspar (a word meaning field crystal) and dark minerals rich in iron and magnesium (ferromagnesian minerals), including black mica—are altered by weathering into fine clay particles.

Black and dark gray crystalline rocks include gabbro (coarse grained) and basalt (fine grained). Because these rocks contain no quartz, soils formed from gabbro and basalt are not sandy but are clayey, sticky, and rather fertile.

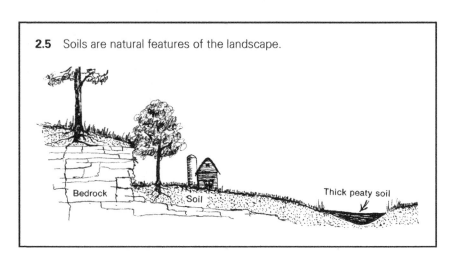

2.5 Soils are natural features of the landscape.

Bedrock

Soil

Thick peaty soil

2.6 Parent material in a topographic location is acted on over time by organisms and climate.

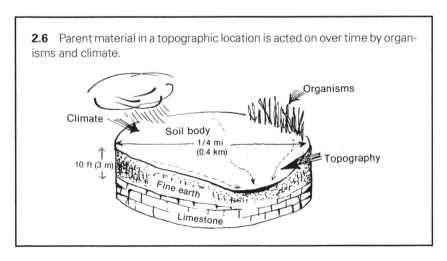

Soils from sandstone are sandy; those from shale are silty or clayey. Soils from limestone consist largely of insoluble shaley materials that were included as gray mud in the otherwise more weatherable rock mass. Therefore, soils from limestone commonly are clayey.

Recent deposits are blankets of geologically young sediments that overlie the types of bedrock just discussed. They include (1) eolian (windblown) sand, (2) loess, (3) volcanic ash, (4) glacial drift, (5) alluvium, (6) and colluvium (Fig. 2.7).

Eolian sands are most common in arid and subhumid areas. Most were initially deposited by water when massive expanses of sandstone were being eroded over a long period. Wind action may shift these cover sands into dune formations, which are then referred to as eolian deposits. The Sand Hills of western Nebraska are a good example of eolian deposits. When viewed from an airplane, they are seen as an expanse of crescent-shaped dunes. They are droughty and not very productive for crops or livestock.

Loess is a wind-transported deposit that mainly consists of silt that was derived from the flood plains of rivers that drained the meltwater from glaciers. These silts have a rich supply of plant nutrient-bearing minerals, and their size is such that they hold a significant quantity of water for crops. Extensive areas of fertile agricultural soils can be found in loess deposits in such places as China, the Mississippi-Missouri Valley, and the Danube Valley in Europe.

Volcanic ash is widespread in Hawaii, Oregon, and Washington in the United States and in Central America, Japan, Indonesia, and many other mountainous areas. The mineralogy of volcanic ash is variable, but most of it develops into high-quality soil for crop production.

Glacial deposits, often with a covering of loess, are parent materials of soils in much of the corn belt in North America and the wheat belt of Eurasia. They were left by glaciers (and their meltwaters) that advanced and retreated repeatedly between 1 million and 10,000 years ago. Glaciers carried a lot of rock debris collected by a grinding action on the terrain over which they passed and thus were made of "dirty" ice. An unsorted mixture (till) of stones, sand, silt, and clay was deposited in broad blankets and ridges called moraines. Glacial till is sometimes stony enough to inhibit cultivation, but its fresh supply of minerals provides an abundance of many plant nutrients. Rapidly flowing meltwaters left behind extensive sheets of sand and gravel, called outwash, that tend to be droughty for crops.

2.7 Bedrock may be blanketed by sediment from several sources.

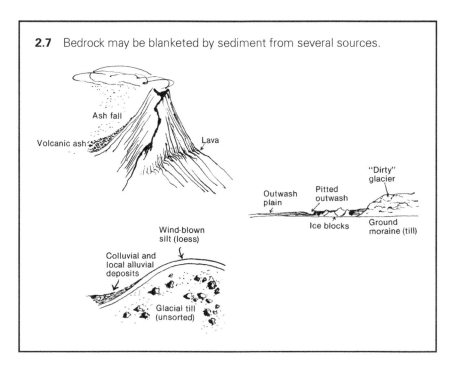

Where huge ice blocks, which melted later, were surrounded by glacial drift (till and outwash), large pits or potholes were formed. Many lakes once existed near the glaciers. Today, the ancient lake bottoms are almost level farmlands with rich silty and clayey soils.

Alluvium is sediment that was deposited by rivers and streams in valleys throughout the world. Centuries of erosion have created fertile areas of alluvial soils: the Bangkok Plain, the Mekong Delta, the Mississippi Delta, and the vast alluvial plains of China. About one-third of the human population is supported on these fertile flood plains that are rich in topsoil materials brought down from the uplands. Although flooding is a major hazard to humans, buildings, and crops, it is a major agent in depositing soil materials. Alluvial soils are finely layered (stratified) to great depths. Each layer may represent the deposit of a single flood. These soils show marked changes horizontally, from somewhat sandy near riverbanks on natural levees and alluvial fans to clayey and even peaty in remote swampy areas. Older soils with distinct subsoil layers may be found on natural terraces, or "high bottoms," that now stand above the rest of the valley floor but were subject to flooding at one time (Fig. 2.8).

Colluvium, a gravity-transported deposit at the base of foothills or mountains, moved from above to its present location. Often, as in the case of mudflows, it was in a somewhat fluid state at the time of transport. These deposits are extremely variable in composition but are not geographically extensive. Colluvium includes talus, which consists of chunks of broken rock at the foot of a mountain.

Climate

Every place on earth has climate that can be described based on its many components. The two components that most strongly influence soil formation are precipitation and temperature.

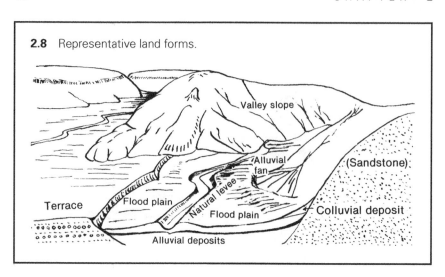

2.8   Representative land forms.

Valley slope

Alluvial fan

(Sandstone)

Terrace

Flood plain

Natural levee

Flood plain

Colluvial deposit

Alluvial deposits

Each of the soil-forming factors interacts with the others, and this is evident with climate. It strongly influences the rate at which rocks are weathered into a loose regolith. It controls the supply of water for physical weathering and determines breakup by freezing and thawing. Climatic change led to the advancement and retreat of glaciers and the resulting glacial till.

It is the effect of climate on chemical weathering that has the greatest influence on the weathering of rocks. Precipitation provides the water necessary for chemical weathering processes and may be sufficient to carry away soluble products, thereby allowing the reaction to continue. Without water, there can be no hydrolysis or hydration. Even oxidation-reduction may be dependent on the quantity of dissolved oxygen. The solution of minerals in certain rocks is dependent on rainfall unless they are adjacent to a body of water.

Temperature has a marked influence on the rate of soil formation. Perhaps the most obvious effect is that which occurs in the temperate zone, where essentially no chemical weathering takes place while the ground is frozen. There is a well-established rule in chemistry that for every 10°C rise in temperature, the rate of chemical reactions increases by a factor of 2–3. For example, the soils of the warmer southern part of the United States are more highly weathered than those in the cooler northern states even where glaciers were not a factor.

The combined influence of precipitation and temperature is probably as important as either one of them individually. If the temperature is cool, water does not evaporate fast, so the effectiveness of the precipitation is high. On the other hand, some warm areas receive quite a lot of precipitation, but due to rapid evaporation, they have the properties of a much drier climate. As an example, St. Paul, Minnesota, and San Antonio, Texas, each receive about 28 in. of precipitation annually, but because of the cool Minnesota temperature, the soil there is normally moist, whereas in the San Antonio area, the soil is usually dry. This effect is also reflected in the native vegetation, which is hardwood forest in the St. Paul area and drought-tolerant vegetation in the prairies of South Texas.

Living Organisms

The influence of all the organisms, plants, and animals (both large and small) is the biotic factor of soil formation. Chapter 4 is devoted to soil biology, but in

2.9 Grass leaves are normally highest in bases, broad leaves of trees are intermediate, and conifer needles are the lowest.

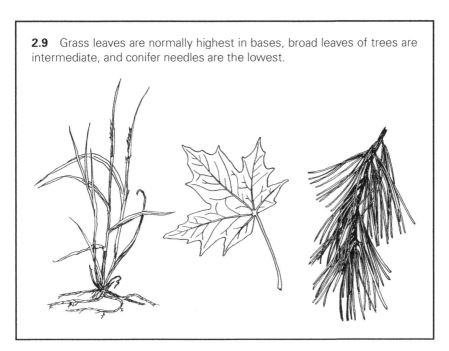

this section the ways that living organisms are involved in soil development are discussed.

In any particular climatic region, the amount of humus in the soil is a direct result of how much and what type of plant residue has been incorporated into it. Thus, if vegetation is sparse, the soil will be low in humus and less fertile. Grasses have a fibrous root system that quite thoroughly invades the tiny pores of the soil so that as the roots live, die, and decay over thousands of years, the soil becomes well supplied with humus. Tree roots are much larger, but because they do not invade the pores of the topsoil as completely as those of grasses, the humus content of soils under forests is usually lower.

Most of the trees in the world's forests can be divided into two groups: the hardwoods with broad leaves and the softwoods (conifers) with needles. Chemical analyses of broad leaves and needles show that needles are usually more acid because they contain fewer base-forming elements such as calcium and magnesium. Grasses contain even more bases than either broad leaves or needles (Fig. 2.9). Therefore, soils formed under conifer forests tend to be the most acid and least well buffered (e.g., against acid rain).

Grassland regions have the most fertile soil for agriculture, but most of them are subject to extended dry periods. Pioneers tended to select the hardwood forests as places to settle because the soils were quite good, and they needed the forest products for their livelihood.

Topography

The lay of the land—its levelness or hilliness—is called **topography.** Topography influences the formation of soil primarily in two ways: (1) Erosion carries topsoil from the higher positions, particularly the side slopes of hills, and deposits it in the valleys.

This results in relatively thicker, more fertile soils in the valleys. (2) Water drains from the uplands to the valleys and if the excess is removed in a timely manner, vegetation is more abundant there. The abundant plant life, which does not decompose as rapidly in moist valleys as on the drier uplands, also contributes to the formation of deep, dark-colored, fertile soils. As a result, much of the world's population relies on crops grown in valleys for their food.

Climatic conditions modify the effects of topography on soil development. In the subhumid and drier climates, the soils are well drained in all positions in the landscape, but they differ in thickness by their long history of erosion or deposition. In the humid regions with a rolling landscape, soils may be thin and excessively drained on the hills and thick with poor drainage in the valleys. Broad, nearly level topographic positions typically have deeply developed soils even if they lie high above the drainageways. In the humid regions, these areas will show the effects of excess moisture unless the parent material is coarse textured, so it will allow rapid internal drainage. In semiarid regions, broad level uplands typically have deep, dark colored soils formed under grassland vegetation.

Topography is a strong indicator of soil characteristics within a particular region. In 1935 an English earth scientist named Milne, working in East Africa, noticed the sequential nature of soils from the top of one hill, down through the valley, up the next hill, and down again repeatedly. Being a scholar in classic languages, Milne knew the Greek term for the arc formed by a chain suspended between two posts. From this, he derived the term *catena,* meaning a sequence of soils differing from each other due to their topographic position. In a two-dimensional sense, he saw each soil as a link in the chain. A catena is a toposequence of soils that may differ from each other in a variety of ways, such as composition and drainage. The drainage catena relationship of a humid region illustrated in Figure 2.10 is horizontally compressed.

Time

Time is typically discussed as the last of the five soil-forming factors. It is a consideration of how long the other factors have been influencing soil formation. The effects of time can best be seen in equatorial regions, where the extremes in age are well expressed. Geologically young areas typically have an irregular topography, and they are comparatively more fertile because young parent materials usually contain an abundance of weatherable minerals that slowly release plant nutrients as they weather. Geologically old surfaces, on the other hand, have long since lost most of their weatherable minerals. Their fertility is found primarily in the organic matter, which is subject to rapid

2.10 In a drainage catena, the soil reflects the effects of long-term moisture conditions.

Brighter subsoil

Gray subsoil

depletion under cultivation. Since prehistoric times, farmers in the tropics have been attracted to rugged landscapes because of the success of growing better crops there. Similar comparisons of soil fertility could be made between geologically young regions such as the northern Rocky Mountains and old, highly weathered portions of the Piedmont Plateau in the southeastern United States.

In glaciated regions, which occur in much of the northern part of the United States, there is a relationship between the time since the last glacial advance, the irregularity of the landscape, and the degree of soil development as evidenced by the concentration of clay

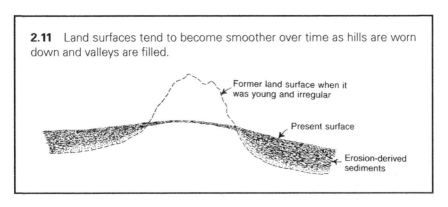

2.11 Land surfaces tend to become smoother over time as hills are worn down and valleys are filled.

in the subsoil. Regions with more recent glacial till (<25,000 years) have many undrained depressions that may form lakes. Moderate to steep slopes are common, and the leaching of clay to the subsoil is moderate. In regions where the glacial till is much older (>50,000 years), more of the depressions have been filled and a complete drainage pattern has formed. The slopes here are more gentle, and there is usually a much greater concentration of clay leached into the subsoil (Fig. 2.11).

Soil Horizon Development

During soil formation both parent materials and organic materials are altered and translocated so that layers called soil horizons develop. The layers usually can be recognized visually. A cross section of soil horizons, called a soil profile, is exposed when a pit or roadside is excavated. Two profiles are illustrated in Figure 2.12. One is typical of some of the subhumid grasslands and the other depicts the soil of humid hardwood forest regions.

Although the number and properties of these horizons vary widely, a rather typical soil profile in a humid region is discussed in this section. Dark humic materials commonly accumulate in the topsoil (the A horizon), followed by a leached zone (E horizon—from the word **eluvial** meaning washed out). The subsoil (B horizon) commonly has an accumulation of clay. The depth to the bottom of the B horizon is typically the depth to which there are abundant plant roots and biological activity. Certainly some roots may extend much deeper.

The portion of the soil profile that has been altered by the soil-forming factors is called the **solum** and is made up of the A, E, and B horizons. On the surface of the A horizon, there may be a layer of plant residue called an O horizon. Below the B horizon the underlying unconsolidated material is called the C horizon. If bedrock is within a few feet of the surface, it is called the R horizon. These symbols may be subdivided with small letters and numbers because of the diverse nature of soil. This system provides symbols used in making detailed soil profile descriptions. The symbols are a type of shorthand used by soil scientists, and they reveal much about the soil properties. The principal soil horizons can be categorized into diagnostic horizons, which will be discussed in Chapter 11.

Leaching of plant nutrients such as potassium and calcium takes place as water moves through the soil, but some nutrients are retained by the finely divided humus and clay materials. Plants take up these nutrients and transport them into their aboveground parts. The nutrients are returned to the soil as the seasons progress; thus, plants contribute to nutrient recycling. This biotic cycling helps to keep the soil from becoming infertile by frequent leaching (Fig. 2.13). Weathering is an ongoing process in the soil and to a lesser extent in

2.12   The profile on the left illustrates a soil from a subhumid grassland; the one on the right shows a soils from a humid hardwood forest region.

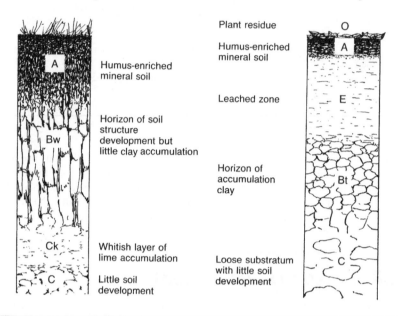

Plant residue

Humus-enriched mineral soil

Humus-enriched mineral soil

Horizon of soil structure development but little clay accumulation

Leached zone

Horizon of accumulation clay

Whitish layer of lime accumulation

Loose substratum with little soil development

Little soil development

2.13   Biotic cycling helps to concentrate nutrients near the soil surface.

CO_2

Nutrients

Nutrients

the substratum below. As soil ages, it is likely to have a higher clay content because clay results from the physical and chemical breakdown of larger particles.

Let's Take a Trip

As we travel from one climatic region to another, there are distinct changes in the native vegetation, and if the farm fields have been plowed, there are differences in the appearance of the soil, even to the casual observer. If the soil is exposed to some depth, there are even more changes evident to those who examine the subsoil carefully.

If we take a trip in the United States from the deserts of the West to the humid woodlands of the East, a succession of soils could be seen (Fig. 2.14). In the arid regions, the tan-colored soil is only a little darker on the surface than it is deeper down because meager rainfall provides for only sparse vegetation. Even here, however, there are differences. Salts may whiten the soil surface in lower areas if water containing large amounts of salts evaporates off the surface. On very old geologic surfaces, carbonates may accumulate in

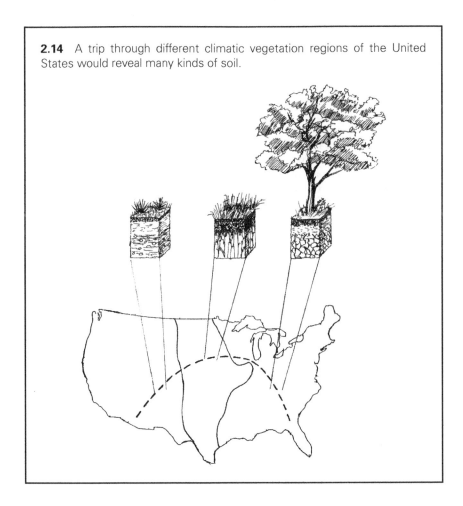

2.14 A trip through different climatic vegetation regions of the United States would reveal many kinds of soil.

the subsoil to form rocklike layers. Pebbles scattered on these ancient surfaces are likely to have a dark reddish-brown varnish from oxides of iron and manganese.

As our trip takes us into the central midwestern states, we enter a region where rain is more common during the growing season, and where the native prairie grasses with their abundant fibrous roots have made the topsoil thick, dark, and rich in plant nutrients. These soils do not have a leached E horizon. They are, in the main, the most productive soils in the United States. When fields are plowed, they appear almost black from the abundant humus, and if a road is cut through them, they show that the humus commonly extends 2 or more feet (61 cm) below the surface.

As the average rainfall and humidity increase toward the eastern one-half of the United States, forests replace the grasslands, and the soils are markedly different. When they are tilled, these soils have a grayish-brown appearance, which reflects their lower content of humus and the presence of a leached E horizon beneath a thin A horizon. The subsoil usually has a concentration of clay that shows up as a reddish-brown horizon in road cuts or other exposures. Many of these soils are very productive, but they require more fertilizer and lime because leaching by greater rainfall has occurred.

If we swing south across the Ohio River, we find soils that are geologically much older, and soils in which the effects of weathering have been greater. Here the cultivated fields are quite red in most places as a result of iron from the minerals that have become oxidized. In these soils, the clay-enriched subsoil forms a much thicker zone, and their native fertility is low.

If you travel from southern Texas to northern Minnesota, you will find that soils in Texas will generally have less organic matter than those in the north. In northern Minnesota the growing season is much shorter due to the cold temperature that slows the breakdown or the organic matter. Through time this results in slower soil carbon cycling and an increase in soil organic matter as you move north in the Northern Hemisphere.

Whenever you have the opportunity to travel, be alert to the change in the soils and landscapes. If you look closely you will see that as the soil changes so do the ways people build roads, houses, and manage the land through tillage and crop selection. You will find that some soils can support a lot people whereas other soils cannot.

CHAPTER 3

Soil Physical Properties

Soil physical properties, those properties that can be seen or felt, are discussed in this chapter. Chemical properties cannot be seen or felt but can be detected with sophisticated scientific instruments. Some chemical properties can be easily altered with soil amendments, but physical properties are often much more difficult to change. Thus, physical properties should receive greater consideration in land-use planning.

Soil Phases

From a physical standpoint, soil is a three-phase system: solid, liquid, and gas. Each phase is equally essential for growth of plants. In a typical soil the solid phase is made up primarily of mineral particles along with a small amount of humus (organic particles). Organic soils commonly found in wetlands may have a high amount of humus particles in addition to the mineral particles.

The solid phase is the source of nutrients and provides anchorage for plants and makes up approximately half of the soil volume. The liquid and gas phases are in the pores between the mineral and organic particles and occupy the other half. The proportion of liquid and gas varies as the soil gains or loses moisture. Plants must be able to absorb water from the soil, and all except a few aquatic plants depend upon the soil pores for the oxygen that is essential for every cell in their roots. Figure 3.1 illustrates the approximate proportion of all three phases in a moist soil.

Soil Separates

The mineral fraction of the soil consists of particles of various sizes. **Soil separates** are mineral particles that are classified on the basis of their size. Sand, silt, and clay make

Soil Science Simplified, Sixth Edition. Neal S. Eash, Thomas J. Sauer, Deb O'Dell, and Evah Odoi.
© 2016 John Wiley & Sons, Inc. Published 2016 by John Wiley & Sons, Inc.

up the soil separates, which are collectively referred to as the "**fine earth**" fraction and are smaller than 2 mm in diameter. The "**coarse earth**" fraction is larger than 2 mm in diameter and consists of gravel, stones, and so forth. The "fine earth" fraction plays a major role in plant growth as well as influences land-use and management decisions.

In the USDA classification system sand particles range in size from 2 to 0.05 mm; silt particles are smaller than sand and range in size from 0.05 to 0.002 mm; and clay particles are smaller (less than 0.002 mm) than silt particles. If the diameter of medium-sized particles of clay, silt, and sand were expanded 1,000 times, the clay would have a diameter about the thickness of this page, the silt about 1 in. (2.5 cm), and the sand about 40 in. (1 m).

Sand

Sand forms the framework of soil and gives it stability when in a mixture with finer particles. Pure sand, however, does not cling together, so it is easily eroded by water and wind. During erosion, sand is not suspended in the water or air but bounces along the surface and piles up where the velocity of wind or water decreases. In the case of wind erosion, this causes sand to form into drifts like snow.

Quartz is usually the dominant mineral in sand because it is the most resistant to weathering of the common minerals in rocks; thus, its breakdown is extremely slow. Many other minerals are found in sand, depending on the rocks from which the sand was derived.

The shape of sand grains is more or less spherical. However, the angularity of sand grains is variable due to the degree to which the specific deposit was rolled around by flowing water.

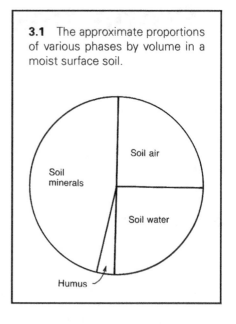

3.1 The approximate proportions of various phases by volume in a moist surface soil.

Sand contributes very little to plant nutrition. The quartz in sand contributes no plant nutrients to the soil while the other minerals, such as feldspars, release their nutrients very slowly. Nevertheless, soils that have a lot of feldspar and other weatherable minerals in their sand fraction develop a comparatively higher state of fertility over the thousands of years of soil formation.

Silt

In many respects, silt is similar to sand except that it is smaller and is too small to be seen with the naked eye. It is spherical and mineralogically similar to sand. Silt is too fine to be gritty to the touch but imparts a smooth feel without stickiness. It is fine enough to be suspended in flowing water, but it drops out when the flow is reduced. This is the reason that harbors are said to become "silted in." If silt is disturbed by drifting sand, it can be picked up and carried great distances by strong winds; thus, silt constitutes the main part of the wind-deposited parent material, **loess**. This concept will be discussed further in Chapter 10.

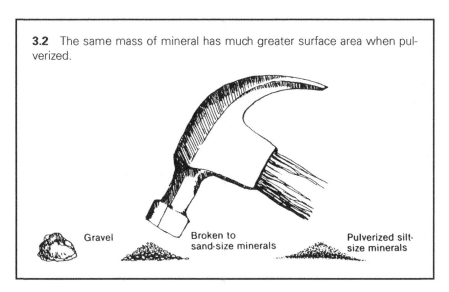

3.2 The same mass of mineral has much greater surface area when pulverized.

Gravel

Broken to sand-size minerals

Pulverized silt-size minerals

Clay

This soil separate is for the most part much different, particularly in size and chemical composition, from sand and silt. Sand and silt are progressively finer and finer pieces of the original crystals in the parent rocks, while clay, on the other hand, is made up of secondary minerals that were formed by the drastic alteration of the original forms or by the recrystallization of the products of their weathering. Clay is so powdery fine that 1 g would have a volume about equal to that of a pencil eraser while the total surface area would equal about one-fifth of a football field (Fig. 3.2). This tremendous surface area results from the platelike shape of the individual clay particles. The maximum diameter of a clay particle is 0.002 mm. Finer clays in the range of 0.0001 mm are called **colloidal clays**. They can only be viewed clearly with an electron microscope.

To illustrate some characteristics of clay, take a large ball of pie dough and roll it into a thin sheet with a rolling pin (Fig. 3.3). Pieces cut from the sheet could be stacked to make

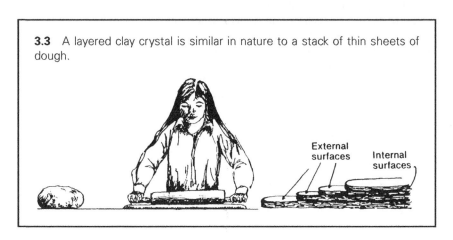

3.3 A layered clay crystal is similar in nature to a stack of thin sheets of dough.

External surfaces

Internal surfaces

a model of a clay particle. The pile of thin sheets would have a much larger surface area, inside and out, than the original ball of dough. Similarly, each clay particle is actually a stack of many very small sheets. There are many kinds of clay, each with different internal arrangements of chemical elements that give them individual characteristics. The major groups of clays related to their chemical characteristics will be discussed in more detail in Chapter 5.

Soil Texture

Soil texture is the degree of fineness or coarseness of the soil. It is an expression of the relative amounts or percentages of sand, silt, and clay. Texture is a permanent property of the soil. In a general way, texture influences the water and nutrient supplying potential for plants; the amount of humus; the volume of pores; the bonding of particles to each other; the ability of the soil to adsorb and hold certain chemicals; drainage of water; and the soil's ability to bear weight. Among the soil separates, clay is the most influential on these soil properties. Many land-use decisions are based on texture of the soil.

All mineral soils can be classified into 12 textural classes of the USDA classification system as represented in the textural triangle (Fig. 3.4). Soils that are dominated by sand are considered "coarse textured," and those dominated by clay are considered "fine textured." Soils that have properties strongly influenced by more than one soil separate are considered "medium textured." Additions of organic matter to a soil (not shown in the triangle) modify soil behavior; sandy soils seem finer textured and clay soils seem coarser textured than they really are. Chemical and biological properties are also changed with the addition of organic material.

Determination of Texture

The proportion of sand, silt, and clay can be accurately determined in the laboratory by measuring the density of a suspension of soil particles in water with a hydrometer. The resulting data (sand, silt, and clay content) are placed on a textural triangle to determine the textural class of a soil. With practice, texture can also be closely estimated by the "feel method," which is commonly used in the field. The "feel method" of texture determination requires rubbing a moist soil between the thumb and the forefinger. Sand in a soil feels gritty. If moist soil feels smooth, but not really sticky, it is a silty soil. If it is very sticky and can be rubbed into a cohesive ribbon that extends from the fingers like a broad blade of grass, it is a clayey soil. Laboratory data can be used to calibrate one's fingers as to the feel of each of the individual soil separates so as to place the soil in the proper textural domain on the textural triangle.

A soil with a significant amount of sand, silt, and clay is called a loam. Various kinds of loams are classified by feel according to the degree of grittiness, smoothness, and stickiness: sandy loam, silt loam, and clay loam. A simple loam without excessive amounts of any soil separate has about 20% clay, 40% silt, and 40% sand (Fig. 3.4). Compared to silt and sand, clay is so sticky that not much is required to give the soil a special texture. Presence of organic matter in a soil (not shown in the triangle) can modify the feel of a soil; sandy soils seem finer textured and clay soils seem coarser textured than they really are.

The texture of a soil does not indicate how it was formed. Did wind, water, or glacial ice drop the particles of sand, silt, and clay at a particular site? Such questions about how the processes of soil formation resulted in sand, silt, or clay fractions were discussed in Chapter 2.

3.4 A textural triangle shows the limits of sand, silt, and clay content of the various texture classes.

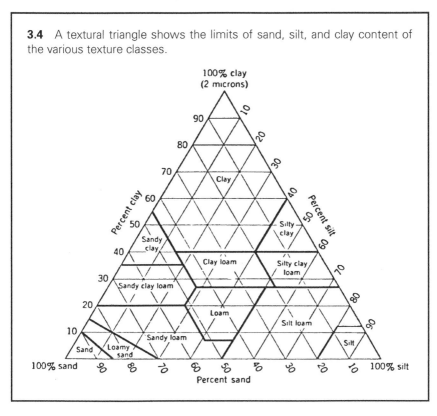

Soil Structure

Individual particles of sand, silt, and clay tend to become clustered into units of various shapes. This clustered unit is referred to as **soil structure**, which is defined as the arrangement of soil particles. The resulting structural units are called aggregates or **peds**. Soil structure creates a range of different-sized pores. Without structure, fine-textured soils would be one massive chunk (with mostly smaller diameter pores) or like loose beach sand (with mostly large diameter pores). Good soil structure means a large volume of pores as well. Good structure allows the soil to retain adequate water as well as drain excess water; promotes ease of seedling emergence, root penetration, and tuber growth; air movement; and erosion control.

Structural arrangements result from biological, chemical, and physical forces that cause the soil particles to bond with each other. Clay and humus because of their small size, high surface area, and electrical charges serve as cementing agents in the bonding of particles within the aggregates. Microorganisms in the soil also play an important role in producing sticky substances that help cement particles together. Oxides of iron and several cations in salts help the bonding process. Physical forces also play a significant role in bringing particles closer together to form aggregates; among these forces are shrinking and swelling from wetting and drying, freezing and thawing, and the actions of expanding roots and of earthworms and other soil organisms. A sandy-textured soil does not have enough cementing agents to hold the soil together as aggregates. As a result these soils are like a sandy beach and are considered "structureless" or "single grained."

3.5 Soil structural units are classified according to shape and size.

Soil structural units (peds)	Soil structural types	Common occurrence
	Granular	In dark surface soil
	Platy	In eluviated subsurface soil
	Blocky	In clay-enriched subsoil, particularly in forested regions
	Prismatic	In subsoil, particularly in grassland regions
	Columnar	In sodium-affected subsoil of grassland regions

As structure is formed, pores are created within and between aggregates. The spaces within the aggregate (between individual soil particles) are small pores or **micropores**, and those between the aggregates are large pores or **macropores**. The larger pores allow water and air to move through the profile, while the small pores act as a sponge and retain water for use by plants. Soil structural units are classified according to shape and size as granular, platy, blocky, prismatic, and columnar structures (Fig. 3.5).

Granular structure is best recognized by farmers and gardeners who strive for a mellow soil. The more or less spherical clusters are called aggregates, and when soil is tilled it can be determined if it is well aggregated by the ease of working it. Some clay and a plentiful amount of organic matter are the key to stable aggregates in the topsoil. The aggregates in coarse-textured topsoil are usually rather porous, like breadcrumbs, and it is described as having a **crumb structure**.

Platy structure has a long horizontal and a short vertical axis. When this occurs in the subsoil, water penetration is restricted. For example, on-site waste disposal systems for rural homes are likely to fail if soil beneath the seepage bed has platy structure.

Blocky structure is the most common structure in the subsoil in humid regions that had forest as its native vegetation. The vertical and horizontal axes are about the same length. This gives a somewhat cubical form that allows good water percolation along the boundaries of the blocks. If there is plentiful clay in the soil, the edges of the blocks are likely to be angular. Structure is less well developed in coarse-textured soils and edges of the blocks are rounded. This is known as subangular blocky structure.

Prismatic structure is best developed in the subsoils with a plentiful amount of clay in regions where the soil becomes periodically desiccated. These conditions are most common where prairie grasses were the native vegetation. The sides of the prisms act as an avenue for water movement.

Columnar structure is an undesirable variation of prismatic wherein the tops of the prisms are rounded and usually covered with gray soil particles. If the topsoil is cleared away, the tops of the columns look like the tops of baking powder biscuits. This happens when there is too much sodium in the soil. This condition is extremely restrictive to water percolation but, fortunately, it is usually localized in semiarid regions.

Soil structural units can also be classified based on the relative bonding strength or adherence of individual particles to each other within a ped. The bonding strength or stability of an aggregate is its ability to resist breakdown from external forces such as raindrop impact and tillage activities. The bonding strength of soil aggregates is classified as **weak**, **moderate**, or **strong**. The aggregate stability of the surface soil is particularly important for minimizing soil erosion.

Benefits of Aggregation

A well-aggregated soil is considered a high-quality soil. A well-aggregated soil is considered to have good tilth because its looseness allows better water infiltration, seedling emergence, root growth, and air and water movement in the root zone.

Aggregates at the surface of the soil are constantly subject to destructive forces (either physical or chemical) that may weaken the bonding or shear the soil particles from the aggregate. Among the physical forces are raindrop impact, rapid wetting, rapid freezing, intensive tillage, compaction due to traffic and harvesting equipment, and so forth. Once the aggregates fall apart, it may take several years of good soil management before the structure will revert to its original state.

Soils under natural vegetation, or heavily mulched soils, tend to have good structure because they are protected from the physical forces of raindrop impact. These soils tend to have more humus than their tilled counterparts, and therefore they are likely to have a healthy population of organisms, which helps protect the soil from falling apart. Without humus, soils with a significant amount of silt and clay become very dense and cloddy when they are tilled repeatedly (Fig. 3.6).

3.6 Soil without humus becomes cloddy (left), whereas humus-rich soil is granular (right).

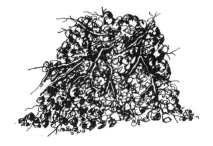

Weakly bonded aggregates at the surface of a bare soil are particularly subject to breakdown by the impact of raindrops. As these aggregates fall apart, the finer particles move into the pores at the soil surface, plugging them and forming a surface crust. In this situation, rainwater will have difficulty entering the soil; instead it begins to flow across the soil surface, creating a potential for accelerated erosion of surface soil particles. Crusted surfaces may also affect seedling emergence. Protecting the bare soil with mulch, crop residue, or vegetation will protect the aggregates from breaking and maintain good structure under the cover. The structure of soils in undisturbed forests is well protected by the canopy as well as a litter layer on the surface.

Excessive weight due to tillage machinery or harvesting equipment can squeeze the soil particles together (minimize the large pores) and compact the soil. The effects of compaction on structure destruction become even more obvious if the machinery is operated when the soil is wet. The water between the particles serves as a lubricant allowing the particles to come closer together due to external pressures. Compaction from the weight of machinery impacts the subsoil's structure and will affect the air–water relations in the subsoil.

Tillage practices that pulverize the soil in the plow layer also destroy the structure. In addition, the churning of the soil exposes humus in the plow layer to air, which increases the rate of breakdown of humus. Consequently medium- and fine-textured soils become dense and cloddy because repeated tillage may have depleted their humus content. For this reason, undisturbed soils such as forest or prairie land tend to have more humus and better structure than their tilled counterparts in the same region.

The goal of good farming, forestry practices, and urban soil management should be to protect soil structure, especially at the soil surface. As part of this goal, managers should avoid soil compaction by minimizing tillage operations, and avoid any tillage activity if the soil moisture is likely to promote compaction. Good management practices must include a plan to increase humus content of the soil by periodically adding organic matter; maintain the humus content at a level that will sustain stable aggregates; plant closely spaced vegetation that has a fibrous root system to restore the humus content and the soil structure; and protect the structure at the surface by not having a bare soil exposed to raindrop impact. If these practices are implemented, natural cycles of freezing and thawing, or wetting and drying, could help restore damaged soil structure over time.

Porosity and Density

The volume occupied by pores in soil is called **porosity**. As the parent material of soil becomes weathered, loosened, and mixed by a variety of forces, pore space develops, providing a place for air and water to be held. Soils that have good structure should have 50% of their volume consisting of pores. Both the amount of pore space and the size of the pores are important. Small pores retain water very well while in large pores, water drains out and air moves in. Sand-textured soils have mostly large pores, and therefore they tend to drain water rapidly. The lack of small pores also makes the sand-textured soils droughty. Clay-textured soils may have a greater proportion of small pores and tend to retain water better and sometimes may become water logged. Therefore, it is desirable to have a balance of both large and small pores such as found in medium-textured soils (silt loams and loams) that are in good structure.

Density indicates the looseness or tightness of a soil. Density of soil, called **bulk density**, includes both the solid particles and the pore spaces among them. If a soil is compacted, the amount of pore space is reduced and the weight of a given volume of soil is increased. The measure of density is a comparison to water, which has a density of 1 g/cm^3. The mineral grains in the soil have a density of about 2.6 g/cm^3. The total volume of the soil is around 40–60% pore space, so by using a mean value of 50% for porosity, bulk density would be 1.3 g/cm^3. This is one-half the density of the minerals in solid rock (Fig. 3.7). Density can be expressed in the imperial system, such as pounds per cubic foot, but it is customary to express density in metric units.

3.7 When rocks weather, they become loosened and less dense as soil is formed.

A core of solid rock

A core of dry soil

Density about 2.6 g/cm^3

Density about 1.3 g/cm^3

Some soils have naturally compacted layers (pans) that may have a high bulk density. Such densities restrict root penetration and water movement. In other cases, heavy tractors and machinery may cause serious compaction (Fig. 3.8), which limits plant growth. In recent years, there has been a shift toward the use of tillage equipment that properly loosens the soil, leaves some protective crop residue on the surface, and allows for fewer trips to be made over the field.

Composition of Soil Pores

Soil pores can be filled totally with air or water. If a medium-textured soil is moist but freely drained, the air and water content of its pores are probably about equal. Normally, soils that seem dry still contain some moisture and the relative humidity in the pores remains near 100%. The water in the pores is actually a soil solution because it contains the ions of dissolved salts. Some are plant nutrients that may be absorbed by plant

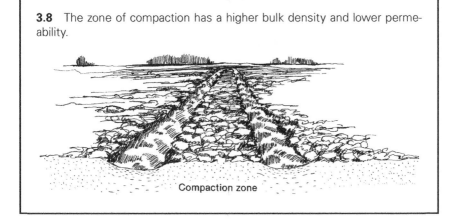

3.8 The zone of compaction has a higher bulk density and lower permeability.

Compaction zone

roots. The soil solution may also contain organic compounds, such as humic and fulvic acids. Humic acid, for example, frequently gives the soil solution a brownish tinge. An abundance of dissolved humus may give alkali (sodic) soils a very dark brown color, but this condition is not widespread.

The Earth's atmosphere is about 78% nitrogen (N_2), 20.9% oxygen (O_2), and 0.03% carbon dioxide (CO_2), with trace amounts of other gases. If the surface soil has free exchange, the soil air and the atmosphere will have about the same composition. However, when the plant roots and soil organisms are flourishing in the growing seasons, CO_2 is being respired by the living cells as oxygen is being absorbed. Nitrogen is essentially inert for all but a few specialized organisms, so its content remains unchanged. O_2 and CO_2 are the main variables. It is common in the root zone for O_2 to drop to 10% and the CO_2 to rise to 10% without ill effects to the plants. Even 5% O_2 and 15% CO_2 may not be harmful, since crops vary in their tolerance to CO_2. When soil pores fill with water, the life-sustaining O_2 is soon depleted. Corn is very sensitive to this condition, but sorghum can withstand several days of flooding without permanent damage.

Soil Consistence

A description of **soil consistence** gives an indication of how soil will react to mechanical manipulation at various moisture contents. The field measurements are made between the fingers, which give a good indication of how the soil will react to tillage, traffic, digging, or similar activity. When the soil is dry, it is described according to a fixed set of parameters as to its degree of hardness or softness. In the moist state, the degree of friability or firmness is used. When wet, it is ranked by its stickiness. The amount and type of clay is the single most important characteristic in determining soil consistence. For example, a clayey soil is likely to be very hard when dry, very firm when moist, and very sticky when wet.

For engineering purposes, more quantitative measurements of soil consistence can be made in a laboratory and expressed as a percentage of water by weight remaining in the soil when the soil displays the following characteristics:

Plastic limit is the moisture content when the soil crumbles as it is rolled into a "wire" between the palm of the hand and a frosted glass plate.

Liquid limit is the moisture content at the point when the soil flows in a curved-bottom dish after 25 impacts in a simple machine that lifts the dish a short distance and lets it drop on a hard surface. A specific tool has been designed for this measurement.

Plasticity index is the difference between the values of plastic limit and liquid limit.

These values are used to predict the relative ease or difficulty of working with earthen materials under differing degrees of wetness.

Soil Color

In Chapter 2, the difference in the appearance of the soil from one region to another was considered. The color changes reflect, for the most part, differences in the quantity of humus and the chemical form of the iron present. It is true, however, that the pigmentation of a given amount of humus is usually darker in grassland regions than in forested regions, particularly in warm areas.

Color of the subsoil gives a strong indication of soil hydrology and the mineral composition of the soil. In some cases, color is an indicator of iron, humus, carbonates, and/or sulfates.

Varying shades of red, yellow, and gray in soils are usually due to the concentration and form of iron present. Red means that the iron is oxidized and not hydrated. Yellow indicates hydration and sometimes less oxidation. Gray indicates chemical reduction caused by wetness and lack of oxygen. An exception to this is the gray E horizon just below the surface of some well-drained soils.

Gray colors in the subsoil or a combination of gray and blotches of yellow and red mottles are extremely important for interpreting the natural drainage condition of the soil. Mottles are found at a depth to which excess water accumulates due to lack of drainage or if the water table rises periodically during the warm seasons. Even when the water table drops, telltale signs of soil colors are left behind, and these are used as a basis for designing septic systems, tile drainage, and the like.

The absorption of solar radiation is greater on dark surfaces than on light ones. This is certainly true for bare soils. Color differences have a comparatively minor effect on the temperature of the soil below the shallow surface layer, but even this can be important for seed germination. Solar radiation has a greater impact on bare soils than on soils with plant cover because when soils become vegetated, leaves intercept the solar radiation before it reaches the soil surface.

Soil scientists use a set of standardized color charts to describe soil colors. These charts are called the Munsell colors. They consider three properties of color—hue, value, and chroma—in combination to come up with a large number of color chips to which soil scientists can compare the color of the soil being investigated. This system is superior to using descriptive terms alone, which may not mean the same thing to everybody.

CHAPTER 4

Soil Biological Properties

Bacteria, fungi, worms, insects, small mammals, and many other organisms inhabit the soil. They participate in and regulate many physical and chemical processes. Soil organisms create favorable conditions for the growth of plants and also decompose plant and animal remains.

Animals in the soil make openings through it that influence the movement of water and air into and through the soil. Termites, for example, air-condition their mounds by channeling air through them. Even the most desolate landscapes on earth have primitive soils, showing the effects of water providing for life in the soil and the translocation of salts and other compounds. There is no soil without life and no higher forms of terrestrial life without soil.

Plant roots can extend down through the soil for several feet (meters). Above ground parts of plants in some forests extend more than 100–200 ft. (30–60 m) high. Shade from the vegetation lowers the amount of soil surface exposed to full sunlight. Roots absorb a large amount of water and this water is conducted through the stems to the leaves where it is either utilized by the plant or it passes into the air as water vapor (see Fig. 6.12). The many tons of plant tissue per acre that die each year—including roots, leaves, fallen branches, and bark—become a part of the soil again through decomposition by soil organisms.

On well-drained uplands, leaves that fall on the forest floor at the end of the growing season in humid temperate regions are nearly all decomposed by the end of the next growing season. In lakes and wetlands, decomposition of plant remains is slowed because the cover of water excludes oxygen. Plant material may accumulate in wetlands as **peat** (which is made up of identifiable plant parts) and **muck** (which is a soil composed of highly rotted, dark organic matter). In upland mineral soils this dark material is called **humus** (Fig. 4.1).

Soils are classified as **mineral** soils and **organic** soils. The difference is in the amount of organic matter present. Arbitrarily, about 25% organic matter by weight is the dividing point between mineral and organic soils. Soils with more organic matter are called organic soils (peat or muck). Soils with less organic matter are called mineral soils because they are composed mostly of inorganic sand, silt, and clay that have been derived from minerals and rocks. A given volume of organic matter is much lighter than an equal volume of

Soil Science Simplified, Sixth Edition. Neal S. Eash, Thomas J. Sauer, Deb O'Dell, and Evah Odoi.
© 2016 John Wiley & Sons, Inc. Published 2016 by John Wiley & Sons, Inc.

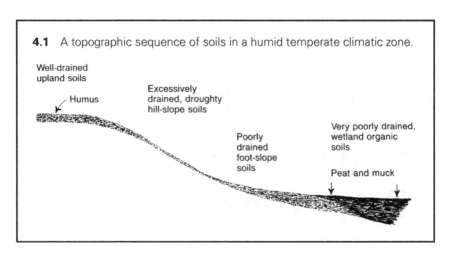

4.1 A topographic sequence of soils in a humid temperate climatic zone.

Well-drained
upland soils

Excessively
drained, droughty
Humus
hill-slope soils

Very poorly drained,
Poorly
wetland organic
drained
soils
foot-slope
soils
Peat and muck

mineral soil. Thus, a soil with 5% organic matter by weight has about 10% organic matter by volume.

Organic Matter and Humus

Organic matter is a general term that includes living and dead organisms, plant and animal residues in various stages of decay, and humus. Soil organic matter (SOM) is usually composed of 50% carbon, 5% nitrogen, 0.5% phosphorus, 0.5% sulfur, 39% oxygen, and 5% hydrogen, but these values can vary from soil to soil. Upland soils consist largely of mineral particles; however, the surface soil, or plow layer, may contain considerable organic matter, which is the partially decomposed residue of plants and animals that live in the soil. Humus gives soil the dark color widely associated with high fertility, although this assumption is not necessarily true for soils that have been heavily cropped or for naturally infertile soils. In most surface soils of temperate humid regions, the humus content is between 1 and 4% by weight (or twice that by volume); but this small quantity has a great influence on the physical, chemical, and biological processes that take place in the soil. Figure 4.2 shows patches of the humus in pores between roots and particles of mineral soil. In arid regions, the surface soil typically has less than 1% humus by weight because temperatures are favorable for organic matter decomposition and vegetative growth is limited by low rainfall.

Humus makes up about 60–80% of SOM and it is derived mainly from plants (flora), with a significant portion coming from the roots (Fig. 4.3), and a very small fraction comes from soil animals (fauna). It is formed by degradation and synthesis processes. In alkaline and neutral soils, the rapid decomposition of plant residues by soil fauna and microorganisms results in the organic fraction of the soil being dominated by humus. In acidic soils, decomposition is slow and plant fragments make a significant contribution to the organic fraction. Soils formed under prairie grasslands generally have greater amounts of humus than do those formed under forest vegetation because of the high density of

4.2 The surface soil contains mineral particles and organic matter.

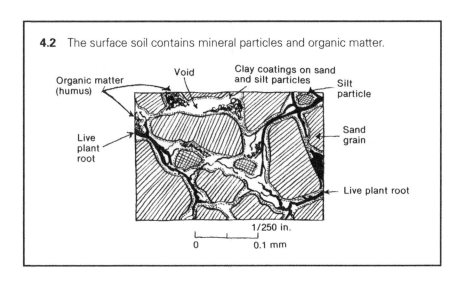

4.3 Humus, shown as a dark layer, can be derived from leaf litter on the forest floor or from roots in surface soil.

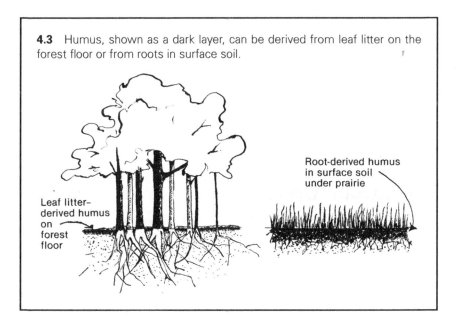

grassland vegetation and the fibrous root system of grasses. In the forest the vegetation at ground level is not nearly so dense, and most of the organic residue from living plants accumulates on the surface of the ground as leaves in the autumn. Much of the humus in the surface, 5 or 6 in. (12 or 15 cm), of forest soils results from incorporation of this plant residue into the soil by insects, worms, and other soil fauna. In agricultural use, the incorporation of plant residue and manure contributes to the formation of humus, but the decomposition of plant roots has been found to be more important.

Humus can be divided into non-humic and humic substances. Non-humic substances consist of carbohydrates, proteins, amino acids, fats, waxes, and low molecular weight organic acids. They are readily attacked by microorganisms and they are rapidly decomposed. Humic substances, on the other hand, are chemically complex organic compounds with large molecular weights and they are therefore relatively resistant to microbial attack. They are mostly responsible for cation exchange and interactions with soil-applied pesticides.

Based on solubility in acid and alkali, humic substances can be further subdivided into:

1. **Fulvic acid**, which is low in molecular weight, light in color, soluble in both acid and alkali, and most susceptible to microbial attack (15–50 years)

2. **Humic acid**, which is medium in molecular weight and color, soluble in alkali but insoluble in acid, and intermediate in susceptibility to degradation by microbes (100+ years)

3. **Humin**, which is high in molecular weight, dark in color, insoluble in both acid and alkali, and most resistant to microbial attack.

The soil is teeming with many forms of life, each occupying a niche that is vital to the entire scheme of life. For microorganisms and small animals, the soil provides environments where conditions of feast or famine may occur side by side or follow each other rapidly. As a moist growing season is succeeded by a dry or cold season, vast numbers of organisms die. Within the soil, small chambers full of rich humus and debris may be separated by volumes of soil that, like underground deserts, are nearly devoid of decomposable organic matter. To survive, therefore, most soil organisms must find something to eat within a few millimeters. Earthworms, on the other hand, are strong enough to make channels and move several feet in search of fallen leaves or other plant debris.

The Carbon Cycle

Life is essential to the existence of a true soil. Of the countless microorganisms that live in the soil, all but a few derive their energy from the oxidation of carbon compounds. Soil organic matter, most of which is humus, serves as an energy source for these organisms. Many functions in the release of nutrients to plants are carried out by soil microorganisms, but only the carbon cycle will be discussed at this point.

During photosynthesis, plants take carbon dioxide (CO_2 from the atmosphere and combine it with water to produce sugar and subsequently all plant tissue. The plants die or are eaten by animals and the residue is returned to the soil. Some of this residue decomposes on the surface while some is incorporated into the soil. Ultimately, most organic material is decomposed by soil organisms and returned to the atmosphere as CO_2, where it can again be used by plants (Fig. 4.4).

4.4 Carbon enters the biosphere through photosynthesis and is cycled back into the atmosphere by decomposers and by burning.

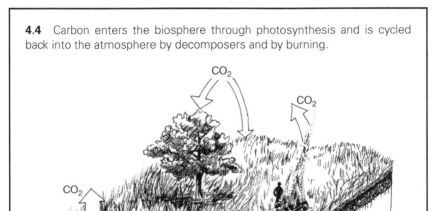

Humus is continuously being decomposed and new humus is being formed to replace the old, except where human mismanagement interrupts the cycle of returning plant or animal residues to the soil, and thus altering (or reducing) the beneficial effects of organic matter. The soil—which supports living plants, animals, and humans—is ever ready to take into itself anything that has died. Any great tree is destined someday to fall and be incorporated into the soil again.

Factors Affecting Soil Organic Matter Levels

Organic matter levels of mineral soils can vary from less than 1% in coarse-textured, sandy soils to more than 5% in fertile, prairie grassland soils. The amount is influenced by the five soil forming factors, discussed in Chapter 2. These factors, arranged in the order of importance, are climate > vegetation > topography = parent material > age.

The following generalizations have been made regarding SOM levels in virgin soils:

1. Soils formed under grasslands generally have greater amounts of humus than soils formed under forest vegetation.

2. The amount of SOM increases with increasing precipitation and decreases with increasing temperature.

3. Fine-textured (clay) soils have higher SOM levels than coarse-textured (sandy) soils.

4. Poorly drained soils have higher SOM levels than well-drained soils.

5. Soils in lowlands have higher SOM contents than soils on upland topographic positions.

The Decomposition Process

During decomposition, complex organic molecules are broken down into smaller and more soluble inorganic molecules such as ammonium (NH_4^+) and nitrate (NO_3^-), among others. The process of transforming organic forms of nutrients into inorganic forms is called mineralization. Soil animals (fauna) perform much of the initial mechanical breakdown of plant residues, after which soil microorganisms (microflora) secrete enzymes (extracellular enzymes) onto the remaining materials and carry out decomposition. Depending on the carbon to nitrogen ratio of the decomposing material, these nutrients may be released into the soil (net mineralization) or they may be used by the soil organisms to build their own cell tissues (immobilization). Soil fauna and microflora also have relatively short life-spans and they are decomposed by other microbes when they die. By breaking down carbon structures and rebuilding new ones, soil organisms play an essential role in nutrient cycling processes and, thus, in the ability of a soil to provide plants with sufficient nutrients.

Mineralization is accomplished by both aerobic and anaerobic organisms using energy derived from carbon contained in the decomposing organic matter. Under aerobic conditions, the main products of decomposition are carbon dioxide (CO_2), water, inorganic nutrients, microbial biomass and humus. In environments where oxygen is in limited supply, the main products of decomposition are methane, some carbon dioxide, hydrogen sulfide, ammonium, organic acids and alcohols. By converting the carbon in organic materials to CO_2, microorganisms complete the biological carbon cycle that was initiated during photosynthesis. Successive decomposition of dead material and modified organic matter results in the formation of a more complex organic matter called humus.

Factors Affecting the Rate of Decomposition

Since soil fauna and microflora are living organisms, they are greatly affected by physical and chemical environmental factors such as moisture, temperature and soil pH. The ideal conditions for decomposition include: moisture content near the soils' water-holding capacity, temperature of 90° to 100°F, soils with oxygen content above 5%, and soil pH near 7.0. Other factors affecting the rate of decomposition are the size of the residue, chemical nature of the organic material as dictated by its nitrogen and lignin contents and C:N ratio, and type and amount of clay minerals present in the soil. Generally, soluble organic materials with simple molecular structures, young leguminous plants, and residues with low C:N ratios tend to decompose most rapidly. Large amounts of clay tend to lower the rate of decomposition.

Importance of Soil Organic Matter

Soil organic matter regulates several attributes that enhance plant productivity and environmental quality.

Soil organic matter is a huge reserve of several nutrients. Mineralization of organic matter by microorganisms releases nutrients (N, P, S, and many minor nutrients) in inorganic forms that can be taken up by plants. Soil organic matter imparts a dark color on the soil, and this may alter soil thermal properties as discussed in Chapter 7. Soil organic matter has the ability to absorb up to 20 times its mass of water, thereby greatly increasing the capacity of soils to store water. The activities of micro and macro organisms promote formation of macropores and aggregates, resulting in improved soil tilth, infiltration/drainage, and reduced erosion. Organic matter complexes with Al^{3+} and metallic ions, particularly Fe^{3+}, Cu^{2+}, Zn^{2+}, and Mn^{2+}, making these micronutrients more available for plant uptake while reducing potential toxicities as well as enhancing the availability of phosphorus in low pH soils. The cation exchange capacity (see discussion on CEC in Chapter 5) of soil humus enhances the retention of cations (e.g., Al^{3+}, Fe^{3+}, Ca^{2+}, Mg^{2+}, NH_4^+), thereby preventing them from leaching to deeper soil layers. Soil organic matter also affects the efficacy of soil-applied herbicides. Other benefits of soil organic matter include increased buffering of soil pH and carbon sequestration.

Carbon Sequestration

One environmentally important function of soil is the sequestration of carbon through plant growth. Sequestration is the taking of gaseous CO_2 from the atmosphere and storing it in stable solid (organic compounds and carbonates) form. It occurs through chemical reactions that convert CO_2 into inorganic carbonates and as plants photosynthesize atmospheric CO_2 into organic compounds in growing plants. A portion of the carbon in plant biomass eventually becomes soil organic carbon during the decomposition process.

The organic matter content of soil usually decreases by between 40 and 60% when grasslands and forests are converted to cropland. This results in the release of CO_2 into the atmosphere. Emission of CO_2 by the burning of fossil fuels and from other sources has been a source of CO_2 increase in the atmosphere. The net effect of increased atmospheric CO_2 is not known, but it has been suggested that reduction of CO_2 in the atmosphere would be "environmentally friendly."

On reasonably fertile soils with reliable water supply, yields in long-term arable agricultural systems have been maintained at very high levels by applying substantial amounts of fertilizer and other soil amendments. In tropical low-input agricultural systems, yields generally decline rapidly as nutrient and soils organic matter levels decline.

The fact that clearing of virgin lands for agricultural use has resulted in losses of a large proportion of soil organic carbon means that there is potential to build soil carbon to improve the productivity of soils. Most SOM is found in the zone of maximum biological activity, that is the topsoil or plow layer. Therefore, anything done to this layer will influence the long-term buildup or depletion of SOM. Since SOM levels are a balance between the rates at which carbon is added and lost from the soil, we can increase SOM levels by increasing carbon input rates and/or decreasing loss rates resulting from decomposition and soil erosion.

Adoption of widespread soil conservation practices has been helpful. Trees are especially beneficial because they sequester carbon in wood for long periods.

Management practices suggested to increase soil organic matter levels include:

1. conservation tillage,
2. proper management of crop residue, such as minimum tillage and stubble mulching, maximizing economic plant populations,
3. application of organic amendments such as manures, composts, biosolids, and Biochar,
4. rotations to include forage or high-residue crops (such as sorghum),
5. precision agriculture, including variable rate application of fertilizer,
6. cover crops,

7. agroforestry in which crops or forage are grown between the rows of trees,

8. pasture establishment using plants with a high proportion of below-ground biomass,

9. irrigation, and

10. terracing.

In summary, any practice is desirable if it decreases decomposition rates and increases yields and/or increases the amount of carbon sequestered from atmospheric CO_2.

Plant Roots and the Rhizosphere

The **rhizosphere** is the volume of soil, water, and air plus associated organisms immediately around the root of a plant. Figure 4.5 shows that the surface of a root is commonly surrounded by gelatinous material in which clay, organic debris, and microorganisms are abundant. Plant roots absorb water and nutrients from the rhizosphere. The roots may release CO_2 and oxygen. The CO_2 makes the soil solution slightly acid so that plant nutrients may be more readily available for uptake. The oxygen may favor precipitation of iron to form a film in the soil near the root. Outer layers of the root may slough off, enriching the soil with organic matter.

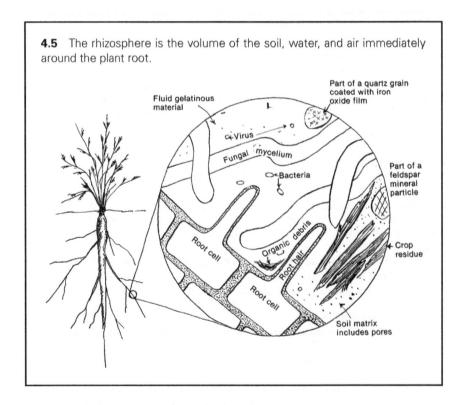

4.5 The rhizosphere is the volume of the soil, water, and air immediately around the plant root.

Microorganisms

Soil microorganisms decompose and dispose of plant and animal remains. In the process, these organisms form humus (Fig. 4.6), which is a more active component of soil than mineral clays. Microorganisms also perform important steps in various nutrient cycles and in solid, liquid, and gaseous phases of the soil–plant root system. Without these organic processes the cycles would lose life support. There is a biological rule stating that the smaller the organism, the greater its number and influence. Thus, the action of microorganisms in the soil is far more widespread and of greater importance than that of insects and rodents.

Some members of each group of organisms perform specialized functions in the soil. It is beyond the scope of this book to discuss all of these, but some are considered in the following paragraphs. Table 4.1 summarizes the essential functions performed by soil organisms.

4.6 Microorganisms in the soil are instrumental in decomposing plant material, resulting in the formation of humus.

Bacteria

Actinomycetes

Fungi

Protozoa

Living organisms are separated into the prokaryotes, which are not clearly either plants or animals, and the eukaryotes, which include the plants and animals. Bacteria and actinomycetes are in the former group; most fungi and all the protozoans are in the latter.

Estimates of the numbers of soil microorganisms in a gram of soil (about the volume of a lima bean) range from several hundred million to a few billion. Most are beneficial to agriculture, but all groups contain those that can cause crop diseases.

Types of Microorganisms

Bacteria are one-celled organisms that are the most abundant forms of life in most soils. They can occur singly or join together in groups. In cropland they are primarily responsible for the decay of residue. They secrete extracellular enzymes that break down organic compounds such as sugars, starches, cellulose, and so on, into basic chemical components like carbon and nitrogen, which the bacteria can use for energy and growth. Any nutrients not needed by the bacteria (or other degrading organisms) are released into the soil and become available for plant uptake. Bacteria also perform a multitude of other functions. Those involved with nitrogen are covered in more detail in the discussion of nitrogen fixation.

Actinomycetes are mycelial bacteria that have threadlike extensions that are all part of the single cell. There may be several million actinomycetes in a gram of prairie soil. They prefer warm, moist soil, and their numbers do not diminish as rapidly with depth as those

Table 4.1   Essential functions performed by soil organisms

Functions	Organisms involved
Maintenance of soil structure	Invertebrates and plant roots, mycorrhizae and other microorganisms that stir or mix the soil
Regulation of water movement in the soil	Plant roots and invertebrates that burrow or bore into the soil
Carbon sequestration and gas exchange	Mostly microorganisms and plant roots, some carbon is protected in large compact aggregates produced by invertebrates
Soil detoxification	Mostly microorganisms
Nutrient cycling	Mostly microorganisms and plant roots, some soil- and litter-feeding invertebrates
Decomposition of organic matter	Various saprophytic and litter-feeding invertebrates (detritivores), fungi, bacteria, actinomycetes and other microorganisms
Suppression of pests, parasites and diseases	Plants, mycorrhizae and other fungi, nematodes, bacteria and various other microorganisms, earthworms, various predators
Sources of food and medicines	Plant roots, various insects (crickets, beetle larvae, ants, termites), earthworms, vertebrates, microorganisms (mostly actinomycetes) and their by-products
Symbiotic and asymbiotic relationships with plants and their roots	Rhizobia, mycorrhizae, actinomycetes, and various other rhizosphere microorganisms, ants
Plant growth control (positive and negative)	Plant roots, rhizobia, mycorrhizae, actinomycetes, pathogens, parasitic nematodes, insects, plant-growth promoting rhizosphere microorganisms, biocontrol agents
Biological decomposition/weathering of rocks	Cyanobacteria, nitrogen- and carbon-fixing bacteria, fungi, and lichens

of other bacteria. The earthy or musty odor of soil comes from the production of geosmin by the actinomycetes. Actinomycetes are important in the degradation of the larger lignin molecules in organic residues. One kind, the streptomycetes, produces antibiotics that we depend on so heavily in many medicines.

Algae are abundant in habitats with adequate moisture and lighting. They can exist as single cells or they can form long chains. Like higher plants, algae contain chlorophyll and are able to convert sunlight into ATP energy and complex organic compounds. Algae frequently live harmoniously or even symbiotically with cyanobacteria (formerly classified as blue-green algae) to form a microbial crust on barren soils. In the United States, these crusts are particularly well developed on arid deserts of the Southwest. Cyanobacteria can fix atmospheric nitrogen (which is discussed under the nitrogen cycle), protect the soil surface from erosion, and create a favorable environment for seed germination. They also fix nitrogen in rice paddy soils and thereby fertilize the growing crops.

Fungi, which are of great importance in decomposing organic residues in the soil, are multicellular organisms ranging in size from microscopic to the large mushrooms normally found only on moist, untilled soil. Fungi common in the soil are made up of a mass of fibers called a mycelium. One hundred thousand fungi may be found in mycelial and spore forms in a gram of soil. They do best in acid soil (pH 4.5–5.5), so they do not compete with bacteria, most of which flourish in nearly neutral soil. Fungi can decompose a greater variety of organic compounds than bacteria. Some catch nematodes in a kind of noose and consume them. Some soil fungi are also pathogenic and cause diseases. The mycelium also function as nets that surround and bind primary soil particles and micro-aggregates into macroaggregates, thereby contributing to soil structural stability.

Mycorrhizae, which means fungus root, is a symbiotic relationship between certain fungi with roots in the surrounding soil. Threads (hyphae) of the mycelia extend into the roots of perhaps half the kinds of higher plants, which means that these plants have "double roots" of high efficiency. The hyphae grow out into the soil and provide water and nutrients, especially phosphorus, for the plants, which in return protect and in part nourish the fungus with sugars. In this symbiotic relationship the fungus may even provide some antibiotic protection to the roots. Mycorrhizae form a sheath around the plant root and either extend the hyphae into the spaces between the root cells or extend the hyphae into the cells of the root where they are finally digested. When the hyphae penetrate the root cells, they form highly branched structures called arbuscules that are the site of nutrient exchange between the plant and fungus.

There are two types of mycorrhizae: endomycorrhizae (hyphae extend into spaces between root cells, but do not enter into the cells) and ectomycorrhizae (hyphae extend into the root cells). The endo group is associated primarily with field crops such as corn, rice, and alfalfa plus a few trees such as apple and citrus. The ecto group is associated mostly with trees, a common one being pine.

Another symbiotic relationship develops between fungi and blue-green algae to form lichens. These primitive plants can survive on bare rock because they fix atmospheric nitrogen and can extract a few nutrients from the minerals of the rock.

Protozoa are one-celled animals. There may be thousands of them in a gram of moist, humic soil. They live inside the films of water that cover soil particles. If the films dry up, the protozoa change into a resting form in which they survive until the next rain. Protozoa include amoeboid, ciliate, and flagellate forms. They contribute to the breakdown of organic matter, and some feed on tremendous numbers of bacteria, thus helping to maintain the balance of nature. It is also thought that by feeding on other soil microbes, protozoa contribute greatly to mineralizing nitrogen in agricultural systems.

Myxomycetes are slime molds, which are intermediate between protozoa and fungi. In the protozoan stage the cells are free-living. In the fungal stage they come together to form a jellylike mass that may be orange, purple, or some other bright color. The fungal stage produces reproductive spores.

The Nitrogen Cycle

Most nutrients, such as phosphorus, calcium, magnesium, and potassium, are derived from minerals. They are absorbed by plants and form living tissue. The plants die and return to the soil, where they decompose and release the nutrients, which can be taken up by plants again. This is a common nutrient cycle.

Nitrogen, however, comes from the atmosphere, which consists of 78% nitrogen in gaseous form. In the nitrogen cycle, nitrogen is transformed from gaseous nitrogen into a form that can be used by plants. Nitrogen undergoes several transformations in the nitrogen cycle. Under certain conditions, the nitrogen returns to the atmosphere before it is utilized by plants. The various steps of the nitrogen cycle are shown in Figure 4.7.

Nitrogen Fixation

Nitrogen fixation is a process that occurs in the nitrogen cycle. It is the process whereby nitrogen from the soil atmosphere is converted into protein in the plant. Nitrogen

4.7   The nitrogen cycle.

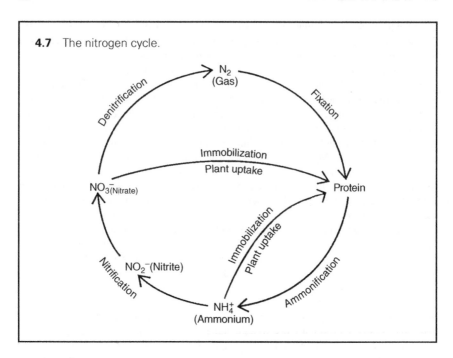

fixation may be **symbiotic** or **nonsymbiotic**. In the case of symbiotic fixation, bacteria live in the root tissue of plants to the mutual benefit (symbiosis) and convert the nitrogen to ammonium (NH_4^+), which can then be utilized by the host plant. The bacteria supply themselves and the host plant with nitrogen, while the host plants supply the bacteria with nutrients and energy sources. Small knots of tissue called nodules form on the roots when these bacteria are present and active. Legumes such as clover, alfalfa, peas, beans, and locust are primary hosts for symbiotic nitrogen-fixing bacteria (rhizobia) (Fig. 4.8).

A vigorous alfalfa crop may fix 100–200 lb of nitrogen per acre (110–220 kg/ha) per year, which is one reason for its inclusion in a crop rotation. Most grasses, including grain crops, are not natural hosts for nitrogen-fixing bacteria. Scientists are trying to find ways of breeding new varieties that can fix nitrogen. If successful, the resultant varieties may reduce the use of commercial nitrogen fertilizer for such crops as corn, wheat, oats, and barley.

Symbiotic nitrogen fixation is also brought about by actinomycetes in association with several woody plants, particularly alder, Russian olive, and sweet fern.

Some nitrogen is fixed nonsymbiotically. A free-living soil bacterium (Fig. 4.8) of the genus *Azotobacter* fixes nitrogen that becomes available to plants when the bacterium dies. The amount of nitrogen fixed in this manner is seldom more than 10 lb per acre (11 kg/ha) per year but is a valuable part of the nitrogen cycle.

Nitrogen fixation also takes place in the atmosphere, particularly by lightning. During storms, lightning will oxidize atmospheric nitrogen to form nitrous oxide (N_2O). It is then carried into the soil by the rain. This is a different type of nitrogen fixation in that the end product is one of the oxides of nitrogen that need not progress through the nitrogen cycle to be available for plant uptake. The amount of nitrogen fixed in this way may average 10 lb per acre (11 kg/ha) per year.

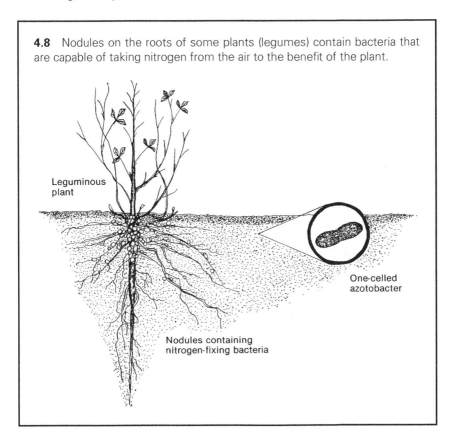

4.8 Nodules on the roots of some plants (legumes) contain bacteria that are capable of taking nitrogen from the air to the benefit of the plant.

Leguminous plant

One-celled azotobacter

Nodules containing nitrogen-fixing bacteria

Ammonification. Ammonification is the first step in mineralization. It is a step in the microbiological decomposition of organic material, such as plant residue, and it is brought about by the general soil population of microorganisms. Ammonia (NH_3) is a product of this decomposition and it ionizes to form the ammonium ion (NH_4^+). Ammonium ions can be held by the soil, fixed in the structure of clay minerals, converted to ammonia and lost to the air via volatilization, absorbed by plants, but most will progress through the next two-part step, nitrification, if the soil is warm and moist. In the case of paddy rice, ammonium ions are the main source of nitrogen.

Nitrification. Nitrification is the reaction that results in the conversion of ammonium ions to nitrate ions. If the soil is warm (75–85°F), has near neutral pH, and is well supplied with moisture and oxygen, the ammonium ions are oxidized first to the nitrite (NO_2^-) form by the bacterium *Nitrosomonas*. The nitrite form rarely accumulates in the soil. The nitrate (NO_3^-) form is brought about directly by the bacterium *Nitrobacter*. These bacteria are able to meet their energy needs by oxidizing NH_4^+ to NO_3^- and obtain C for building their cell structures from carbon dioxide. Nitrate is the highest oxidation state for nitrogen. Since nitrate is an anion, it is not held on the exchanges sites on soil colloids. This, coupled with its high solubility in water makes nitrate subject to leaching, particularly in coarse textured soils, excessively irrigated fields and in areas with high rainfall intensity. If water percolates through the soil, nitrate moves with it and may contaminate groundwater.

Excessive nitrate in drinking water interferes with bloods' ability to carry oxygen resulting in a condition known as Blue Baby Syndrome in infants under 6 months of age. For this reason, the United States Environmental Protection Agency has set 10 milligrams of nitrate nitrogen per liter of water as the maximum concentration safe for drinking water. Surface runoff of nitrate causes eutrophication, or algal blooms, in lakes and estuaries. Eutrophication is the slow, natural nutrient enrichment of streams, lakes and reservoirs. As the algae grow and then decompose, they deplete the water off the dissolved oxygen, resulting in fish kills, offensive odors, and reduced attractiveness of the water for recreation and other public uses.

During the process of nitrification, hydrogen ions are released. The ammonium in fertilizers also undergoes this same nitrification process. This explains why soil acidification occurs when large amounts of organic materials or ammonium containing fertilizers are added to the soil (see Chapter 5).

Immobilization and Mineralization

Immobilization is the conversion of inorganic, primarily ammonium and nitrate, nitrogen into organic nitrogen (amino acids and proteins in microorganisms and plants). Mineralization is the reverse process of immobilization, wherein organic forms of nutrients in organic materials are converted to the inorganic forms by soil organisms during decomposition.

The addition of a large amount of residues with inadequate amounts of nitrogen, such as wheat straw, to the soil stimulates the growth of a large population of microorganisms. After ammonification and nitrification, the ammonium and nitrate ions may be taken up by the roots of higher plants or by microorganisms decomposing organic residues in the soil. Immobilization is the process during which an overabundance of microorganisms, which also need nitrogen to live, may outcompete with crops for the available nitrate. As inorganic ammonium and nitrate are incorporated into the cells of living microorganisms, the plant-available N levels in the soil are reduced. As a result, crops may become nitrogen deficient and develop a yellow coloration. This situation can be prevented by compositing residues before incorporation in the soil, blending low C:N ratio residues with high C:N ratio residues or by adding inorganic nitrogen fertilizers to the soil.

Mineralization and immobilization can have a major influence on the amount of available N in the soil. As has been discussed, soil microorganisms will break down plant material (and other organic materials) to obtain carbon and energy. During the initial stages of decomposition, N can be released (mineralized) from the organic material and it can be taken up (immobilized) by the biomass. The net result of these N transformations will dictate whether available N increases during the early stages of decomposition or decreases.

The amount of N mineralized is proportional to the quantity of total N in the substrate being decomposed. In general, 2% N is considered critical, with net mineralization occurring when the proportion of N in the material is >2%. The relative amount of C and N (C:N ratio) in the decomposing material also determines whether mineralization or immobilization predominates. Since most plants are about 40% C, the C:N ratio is primarily influenced by N content. Plants with higher N contents have lower C:N ratios, and mineralization (increase in available N) is favored when these plant materials decompose. For example, when alfalfa decomposes, the relatively high amount of N (C:N ratio = 25:1) favors mineralization and available N will increase. If residues with lower N contents, and which have higher C:N ratio, are added to the soil, the microbes will have to scavenge the soil for nitrogen to balance the excess carbon in the residue and immobilization (decrease in available N) will usually occur when these plant materials decompose. For example, when corn (C:N ratio ~ 57:1) or wheat (C:N ratio ~ 80:1) residue are added to the soil, the relatively low amount of N (high C:N ratio) favors immobilization and available N decreases. This is because the additional N required to balance the high C in the straw will have to come from the soil. For this reason, it is advisable to add 15 lb of nitrogen for every ton of straw up to 50 lb of N.

The processes of mineralization and immobilization occur simultaneously. As organic matter decomposes, inorganic nitrogen is released into the soil and is utilized by both plants and microorganisms. The N in both the plants and the body mass organisms eventually reverts into plant-available N when they die and decompose to release inorganic nitrogen into the soil through mineralization.

In summary, everything else being equal, mineralization and immobilization proceed at fairly equal rates when the C:N ratio of decomposing organic residues is between 24:1 and 30:1. This implies that microorganisms need 1 gram of N for every 24 g of C in the substrate (food). Net mineralization occurs at C:N ratios below 24:1 while net immobilization occurs at C:N ratios above 30:1.

Crop residues on the soil surface serve to protect the soil from the destructive impact of raindrops, thereby protecting the soil from erosion. The faster the rate of decomposition, the less the time those residues will be available to provide cover to the soil surface. Therefore, while the decomposition of crop residues is important for nutrient cycling, it is also essential to maintain a certain amount of residue for soil cover. For these reasons, it is important to pay attention to the C:N ratio of crop residues, so as to maintain soil cover when desired, yet allow the residue to ultimately break down and be recycled.

Denitrification

Denitrification is the process whereby nitrate nitrogen (NO_3^-) undergoes chemical reduction into gaseous nitrogen forms including nitric oxide (NO), nitrous oxide (N_2O) and molecular nitrogen (N_2), and is volatilized into the atmosphere. This is a process whereby microorganisms that flourish, under anaerobic conditions, derive their oxygen from the nitrate ions or similar oxides. Therefore, for denitrification to occur carbon and nitrate must be available and oxygen availability must be restricted. Most NO_3-N may be lost in this manner from low, wet areas of a field, especially during periods of warm weather and heavy rainfall when the soil stays saturated for prolonged periods of time. Even in aerated soils, small, localized areas in the soil (microsites) can have inadequate supply of oxygen. The presence of crop residues and other decomposable organic matter increases the rate of denitrification. For instance, the presence of surface mulch and moisture in no-till systems greatly increases the potential for denitrification loss of surface-applied N and a side dressing of N fertilizer may become necessary to ensure an adequate supply of N during the early growth phase. Nitrate losses due to denitrification can be reduced by applying slow release fertilizers or fertilizers that contain nitrification inhibitors. Nitrification inhibitors slow the processes of nitrification until periods of greater plant uptake.

To a farmer, denitrification can be a costly loss of an expensive plant nutrient—nitrogen. This is a major reason for maintaining adequate soil drainage and proper timing of nitrogen fertilizer application. The N_2O released into the atmosphere during denitrification is a greenhouse gas that contributes to global warming. Nevertheless, denitrification is a crucial part of the nitrogen cycle as it is the only point in the cycle at which fixed nitrogen re-enters the atmosphere as gaseous nitrogen. Without it, atmospheric nitrogen would eventually be depleted by the nitrogen fixers.

Volatilization

This is the production of gaseous ammonia from ammonium and its loss to the atmosphere. Ammonia volatilization increases with soil pH because a high concentration of OH^+ ions promotes the conversion of nitrate to ammonium. As discussed in Chapter 8, volatilization losses are high for broadcast unincorporated urea fertilizer or manure. Incorporation of manure and fertilizers can reduce ammonia losses by up to 75%. Evaporation promotes volatilization. Thus, volatilization is greatest as the soil dries after reaching field capacity. Crop residues that are not incorporated into the soil may increase the rate of volatilization. Volatilization losses can be reduced by applying slow release fertilizers or by using fertilizers that contain urease inhibitors. Slow release fertilizers contain a coat of sulfur that must break down before urea is released while urease inhibitors slow the process of urea hydrolysis.

Ammonium Fixation

This is the trapping of ammonium ions between interlayer spacing of some 2:1 layer silicate clay minerals such as vermiculite and illite. It occurs because the size of cavity left by oxygen in clays is sufficiently large to hold potassium and ammonium ions, but too small for other ions. Whether ammonium will be fixed or not is determined by the source of charge on clays. Kaolinite does not fix ammonium. Hydrous oxides at low soil pH values do not fix ammonium or potassium because the Al(OH) in the interlayer spaces satisfy the clays' charge and also expands the interlayer space, thereby impeding fixation. Montmorillonite does not fix ammonium under wet soil conditions.

The fixation of NH_4^+ ions leads to a temporary immobilization of fertilizer N applied in a soil. The actual amount of ammonium fixed depends on the amount of K^+ in the fixed position. The more the quantity of K fixed, the less the quantity of ammonium can be fixed. For this reason, NH_4^+ fixation can be reduced by K fertilization prior to NH_4^+ application.

Biological Decomposition of Rocks

Three years after the island of Krakatoa was largely blown away by a violent volcanic eruption in 1883, scientists visited it only several years later and found that the surface of the fresh bedrock was already being invaded by cyanobacteria, one of the most self-supporting forms of life on earth. It can both photosynthesize and fix nitrogen. Growing along with the cyanobacteria were nitrogen- and carbon-fixing bacteria as well as fungi and lichens. Weak acids produced by these microorganisms were dissolving nutrients (phosphorus, calcium, and other nutrients) from the rocks and building up a humic mat capable of supporting mosses and eventually higher plants. The weak acids include carbonic acid formed by solution of CO_2 gas in water and lactic acid produced by fungi, and stronger acids (nitric and sulfuric) that were formed by bacteria. Certain fungi and bacteria can release phosphorus from mineral particles. It is evident that microorganisms are involved in rock weathering from the start.

Macroorganisms

Macroorganisms include worms, arthropods, and vertebrates. In an acre of soil there may be a million nematodes, a million ants, two hundred thousand mites, and four thousand worms, to name just a few. Most wild bees nest in the soil and in the process make the soil more porous by excavating burrows and chambers. Before settlement by European immigrants, a squirrel could cross the state of Ohio without touching the ground. Many of the trees in the native forests were planted by squirrels. Obviously, animal life has greatly influenced both plants and soil.

Worms include nematodes and earthworms. Nematodes are eel-shaped, unsegmented, colorless worms and are generally the most abundant multicellular organisms in soils. (Fig. 4.9). Most are too small to be seen without a microscope, but some may grow to a centimeter or more in length. Many are saprophytic, which means that they feed on dead plant residue, but some are parasitic and live on the roots of plants. Many of the parasitic species cause important diseases of plants, animals, and humans. They cause great economic loss to many crops, including citrus, cotton, soybeans,

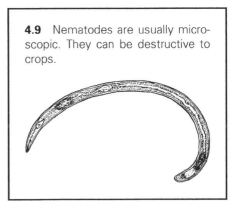

4.9 Nematodes are usually microscopic. They can be destructive to crops.

alfalfa, corn, and vegetables such as potatoes and tomatoes. Nematodes are involved in organic matter decomposition and nutrient cycling, biological control of insect pests and certain plant parasitic nematodes. They also serve as food for other soil organisms.

Earthworms (Fig. 4.10) perform an important function in mixing organic matter with mineral matter. In a sense, they are soil factories. Among the many kinds of earthworms, the nightcrawler, *Lumbricus terrestris*, was brought to the United States from Europe by settlers.

In general, worms perform an important aeration and mixing function by burrowing/channeling through the soil, consuming organic matter, and bringing the residue to the surface as castings, which form stable aggregates upon excretion. It is estimated that worms bring 7–18 tons of soil per acre (16–40 t/ha) annually to the surface in this way.

Arthropods include springtails, mites, and ants. Springtails (*Collembola*) are primitive insects that do not go through stages of metamorphosis as do flies and butterflies (Fig. 4.11). They look like ancient fossil creatures. They are numerous in decaying leaves, and in late winter they appear on snowbanks (hence their nickname "snow fleas") where they feed on scattered pollen.

Mites (*Acarina*) (Fig. 4.11) perform the same job as springtails, which is to consume dead and decomposing plant parts. Mites are found everywhere, even in ocean depths and on high mountains. They consume organic residues and feed on nematodes and springtails.

In both urban and rural environments, **ants** are active in tunneling and bringing up subsurface soil to construct mounds of various sizes as shown in the figure. Because ants can carry particles no larger than allowed by the gap between their open mandibles (mouthparts), the mounds contain no stones or gravel. Figure 4.12 shows a cross section through a mound nearly 1 ft (30 cm) high that was built by the western mound-building ant, *Formica cinera*. These insects were originally common in the grasslands of the American prairie but are now confined by cultivation to undisturbed lands such as those along

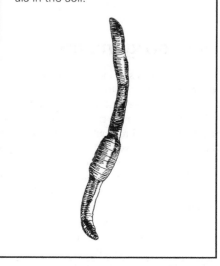

4.10 Earthworms are essential for mixing organic material with minerals in the soil.

railroad tracks, cemetery edges, and wetland borders. Their mounds are built largely of subsoil and are rich in organic materials that the ants bring to the colonies from nearby vegetation.

Termites are particularly active in soil and plant materials in subtropical and tropical regions. They consume large quantities of dead trees, shrubs and plant debris. Some of these insects tend to concentrate nutrients such as calcium in their nests, which, when abandoned, are cultivated by farmers and eventually produce patches of high-quality crops. In semiarid regions, underground termite nests may act as sinks (collectors) for irrigation waters and thereby become a nuisance. The long-term soil-mixing effects of termites are beneficial, but the immediate effects may be troublesome. Some mounds may be higher than those shown in Figure 4.13.

Vertebrates include moles, mice, ground hogs, and many others. Moles (*Talpidae*) plow soil by burrowing just below the surface to where they can find earthworms, grubs, and plant roots to eat. This activity occurs both in sod and in forest topsoil. It leaves the soil loosened and contributes to the high porosity of noncultivated soils.

4.11 Springtails and mites play an important role in the decomposition of dead leaves and stems.

Springtail (Collembola)

Mite (Acarina)

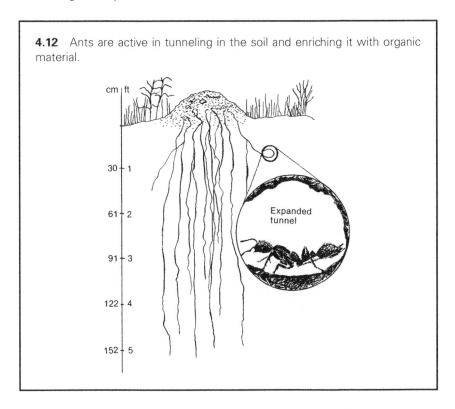

4.12 Ants are active in tunneling in the soil and enriching it with organic material.

Expanded tunnel

Mice (*Cricetidae*) and **shrews** (*Soricidae*) are numerous enough to make an impact on soils by their burrowing activities. When snow melts in the spring, networks of rodent runways are plainly visible. **Ground squirrels** (*Spermophilus*), **Ground hogs** (*Marmota*), **prairie dogs** (*Cynomys*), and other mammals make elaborate burrows, constructed to not fill with water readily during rainy periods and to be aerated by convection and updraft air currents. These rodents bring tons of subsoil material to the surface. Because these animals prefer dry sites, the materials they excavate are commonly sandy and gravelly resulting in a soil profile that is mixed with various soil particles, but enriched with vegetative debris and rodent excreta in the process (Fig. 4.14).

Pesticide Use and Soil Organisms

Before the dawn of agriculture, all organisms were in balance and none were able to build up in numbers beyond that of natural populations. This is not to say that primitive humans were not bothered by insects and the like, but they were natural populations. When humans began to manipulate plants and animals to increase their food supply, the balance was altered so that certain organisms became detrimental to agricultural production. Various forms of control have been used, but in recent decades the emphasis has been upon organic compounds that are intended for the selective control of specific target organisms. Some pesticides kill certain kinds of pests such as fungi, nematodes, insects, or rodents; some regulate plant growth by speeding it up or retarding it; some defoliate or desiccate plants;

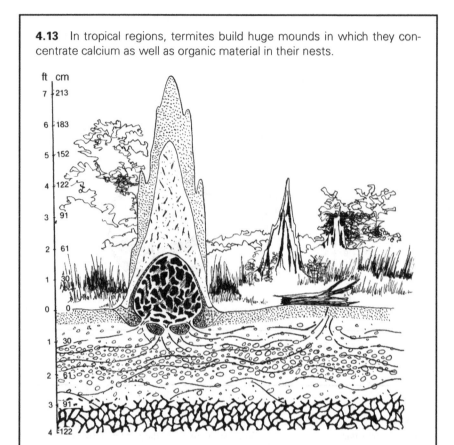

4.13 In tropical regions, termites build huge mounds in which they concentrate calcium as well as organic material in their nests.

some attract insects to deadly traps or sterilize them; and some repel pests through protective coatings such as are found on some seeds.

When properly handled, pesticides cause little or no problem in the environment. One reason for this is the action of soil microorganisms. Bacteria quickly break down most pesticides into components that are harmless when deactivated. Components that are unlike any natural molecules, however, cannot be attacked by bacteria and they can build up in the environment, such as DDT (dichlorodiphenyltrichloroethane).

There are other examples where a component formed by microbial breakdown of the original insecticide is potentially very hazardous to human health. One of the aldicarbs derived from a combination insecticide–nematocide used on potato fields is in this category. Aldicarb (or its derivatives) has been found in the groundwater beneath sandy soils in Wisconsin, New York, and Florida. If the soil is not rapidly permeable, the breakdown of aldicarb into harmless components seems to be complete.

The massive use of pesticides over large areas has, in some cases, been self-defeating. Sometimes natural enemies of a pest have been eliminated, and pesticide-resistant varieties of the pests have evolved. Well-planned harmonizing of chemical and natural control methods (integrated pest management) is a wiser approach.

4.14 The burrowing activities of animals contribute to the porosity and enrichment of soils.

Limited strategic use of pesticides may be combined with ecological pest control. The latter includes the encouragement of growth of populations of natural enemies of pests, release of many sterile individuals of a species, and rotation of crops in a way so as to interrupt population expansion. The tobacco hornworm moth, for example, has been controlled by light use of pesticides together with a vigorous encouragement of parasitic wasps (biological control) and some handpicking (scouting) of larvae. Integrated pest management has been well received because it is an economically sound approach as well as being good for the environment.

The recent development of hybrids that are resistant to specific insects and infections also offers an opportunity to reduce the application of pesticides. Some examples are corn that is resistant to corn borers, potatoes that are resistant to potato beetles, and alfalfa that is resistant to leaf hoppers. The application of insecticides to control these and other insects has been extremely expensive and controversial.

CHAPTER 5

Soil Chemical Properties

The chemically active fractions of the soil are clay and humus. Both clay and humus have electrically charged sites on their surfaces—both negative and positive. These sites attract ions of the opposite charge. The types and relative amounts of ions that are attached will influence the plant nutrient level as well as the alkalinity or acidity of the soil.

Soil Colloidal System

The humus and clay fractions are often called the colloidal system. A colloid, by definition, is an extremely small particle. Clay and humus fit the definition. Colloids are too small to be seen with a light microscope, but clear images of them can be made with electron microscopes. The upper limit of their diameter is commonly given as 0.0001 mm, although particles somewhat larger may react similarly but to a lesser extent. For comparison, it would take 254,000 of these particles, side by side, to extend 1 in. (2.54 cm). Thus, the colloidal system is made up of the finest clay particles and highly decomposed humus. Due to their small size, colloids have a large specific surface area and carry an electrical charge on their surface. As a result, colloids are the most chemically active fraction of the soil and are intimately associated with many reactions involved in plant nutrition.

Since colloidal clay and humus particles have negatively and positively charged sites, nutrient ions that are essential for plant growth are attracted to the colloidal surfaces of opposite charge. The positively charged ions are **cations** and those with negative charge are **anions**. They are held weakly as a reserve supply for plants and may be released into the soil solution where they can be utilized by plants. Without the attraction between ions and colloids, the leaching of certain ions deeper into the soil and beyond the reach of roots would be much greater in humid regions. Indeed, it is often observed that nitrate leaches readily in soils in humid regions. Nitrate ions are negatively charged and are not attracted to the negatively charged soil colloids; therefore, nitrate ions remain in the soil solution.

Soil Science Simplified, Sixth Edition. Neal S. Eash, Thomas J. Sauer, Deb O'Dell, and Evah Odoi.
© 2016 John Wiley & Sons, Inc. Published 2016 by John Wiley & Sons, Inc.

Since the nitrate ions are quite soluble and are not prone to other sorption reactions, nitrate will readily leach.

The nature of the colloidal system is not only dependent on the colloids themselves but also on the properties of the ions attracted to them. These attracted ions may be exchanged, partly in accordance with the dominance of specific ions in the soil solution. This process is called ion exchange. In all soils, except some in tropical regions, the negatively charged sites on colloidal surfaces are much more numerous than are the positive sites, so the usual process is cation exchange.

To understand how colloids influence soil chemistry, it is necessary to know something about their composition. Clay mineral colloids, primarily silicate clays, oxide clays, and humus colloids, will be discussed separately.

Silicate Clays

Mineral particles such as common feldspar grains from granite are made up mostly of three elements: silicon, oxygen, and aluminum. Therefore, they are called aluminosilicates. Small feldspar particles slowly change to clay minerals by weathering. These are also aluminosilicates, but they are different from feldspars in two principal ways: The clay minerals have some water molecules in their structure so they are called hydrated aluminosilicates, and they have a platy or layered structure.

Just as a plant leaf is made up of distinct layers of cells, the very small, flat clay crystals are made up of definite layers of ions. Most silicate clay particles are sandwich-like, with an alumina layer (aluminum plus oxygen) sandwiched between two silica layers (silicon plus oxygen). They are called 2:1 clays because of this arrangement. Smectite and hydrous mica are clays of this type.

In clay minerals with a 1:1 structure, there is a single silica layer adjacent to a single alumina layer. Kaolinite is a common 1:1 clay. Plates of halloysite, a variety of kaolinite, tend to curl (Fig. 5.1).

These 2:1 and 1:1 types of clays are called layer **lattice silicate** clays. The ions in each layer are arranged in lattice-like geometric patterns (Fig. 5.2). The 2:1 lattice clays

5.1 Clay particles are extremely small and in some types the layers tend to curl.

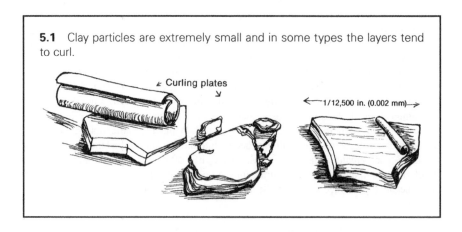

Curling plates

←—1/12,500 in. (0.002 mm)—→

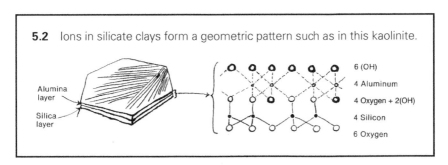

5.2 Ions in silicate clays form a geometric pattern such as in this kaolinite.

Alumina layer

Silica layer

6 (OH)

4 Aluminum

4 Oxygen + 2(OH)

4 Silicon

6 Oxygen

have variations within the geometric pattern of ions that give rise to a negative charge on the surface. Most 2:1 clays are also expanding lattice clays so they absorb water between, but not within, the sets of 2:1 lattices.

As an analogy, clay particles resemble a stack of sandwiches and the expansion takes place between the sandwiches. Expanding lattice clays have a tremendous surface area because the internal surfaces are available to react with the soil solution. Clays with a 1:1 lattice do not expand because hydrogen bonding between the sets of lattices holds them together.

Another kind of clay is the oxide clay that has little or no regularity in its structure. In this respect, oxide clays are gel-like.

Source of Negative Charge on Silicate Clay Minerals

The negative charge on silicate clays in soils comes from two sources. A typical silicate clay of the 2:1 type illustrates this principle. First, the silica layer develops a negative charge from the oxygen ions along the edge of the crystal. Only one of the oxygen's two negative charges is combined with a silicon ion, so at the plane where the crystal ends, there are oxygen ions with one negative charge unsatisfied. Figure 5.3 depicts this charge

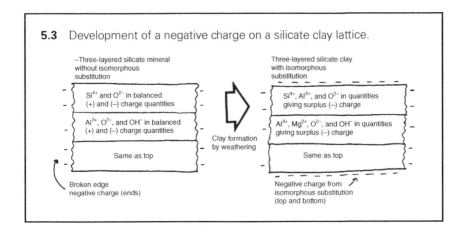

5.3 Development of a negative charge on a silicate clay lattice.

−Three-layered silicate mineral without isomorphous substitution

Three-layered silicate clay with isomorphous substitution

Si^{4+} and O^{2-} in balanced (+) and (−) charge quantities

Al^{3+}, O^{2-}, and OH^- in balanced (+) and (−) charge quantities

Same as top

Clay formation by weathering

Si^{4+}, Al^{3+}, and O^{2-} in quantities giving surplus (−) charge

Al^{3+}, Mg^{2+}, O^{2-}, and OH^- in quantities giving surplus (−) charge

Same as top

Broken edge negative charge (ends)

Negative charge from isomorphous substitution (top and bottom)

distribution in two dimensions, with an unsatisfied charge at each end of the lattice. The oxygen ions are not shown in this schematic diagram, but their location is similar to the ionic arrangement in the silica layer shown in Figure 5.2. This source of negative charge is called edge charge and although it is low, it is the main charge on kaolinite clay, which is a silicate clay mineral. This charge fluctuates with soil pH, hence it is called pH-dependent charge.

The second source of a negative charge arises when one ion is substituted for another during the formation of the silicate clay crystal, without any change in its form. During substitution, some atoms in the crystal are replaced by other atoms of similar size but different valence. This is called **isomorphous** (*Iso* = similar, *morphous* = size and shape) **substitution**, and it can occur in different ways. In some clays an aluminum ion (Al^{3+}) substitutes for a silicon ion (Si^{4+}) in the outer (silica) layers, whereas in other clays a magnesium ion (Mg^{2+}) may substitute for Al^{3+} in the alumina layer. Either way, one negative charge results in the crystal and the charge is permanent since it does not vary with soil pH. In essence, the substitutions result in a deficit of positive charges, and this results in an overall net negative charge on the clay.

Groups of Silicate Clays

Several groups of layer silicate clay minerals have been identified and within each group there are many specific clay minerals. In this book, only three of these groups are discussed to illustrate the nature and importance of clay.

Smectite Group. **Montmorillonite** is a common member of the smectite group. It is a 2:1-type clay with a high capacity to hold plant nutrients and to swell and shrink on wetting and drying (Fig. 5.4). Variations within this group are due mainly to the amount of substitution of magnesium and ferrous iron for aluminum in the alumina layer. Soils that have high amounts of montmorillonite clay can be very troublesome, particularly when wet. They are expanding lattice clays wherein their strong affinity for water causes the clay

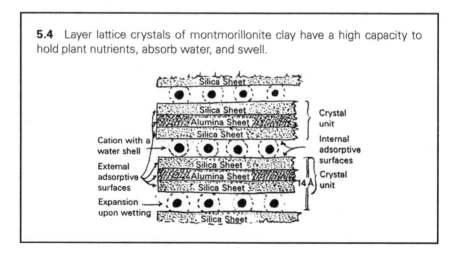

5.4 Layer lattice crystals of montmorillonite clay have a high capacity to hold plant nutrients, absorb water, and swell.

particles to spread apart and readily slip past one another. This results in what is called low bearing strength, which means that foundations of buildings and roads built on these clays are likely to fail (slip) and cause cracking in the superstructure, particularly on sloping ground. When montmorillonitic soils dry, cracks of nearly 2 in. (5 cm) or more may open. Debris may fall into these cracks and cause the soil to buckle when it is wetted.

Montmorillonitic soils become very sticky and difficult to till when wet and very hard when dry. As a result, farmers can work them only at just the right moisture content. Unimproved roads on montmorillonitic soils become impassable in rainy seasons.

The inner (alumina) layer of the montmorillonite clay lattice is made up of aluminum, hydrogen, and oxygen ions. All the negative and positive charges balance and neutralize each other within this layer only if the three named ions are present. In montmorillonite clay, about one-fourth of the aluminum ions (Al^{3+}) have been replaced by ions of magnesium (Mg^{2+}) or iron (Fe^{2+}); ions with two positive charges have been substituted for ions with three positive charges. This produces a deficiency in positive charges, which results in an excess of negative charges at the surface of the crystal lattice. These are permanent negative charges that developed when the crystals were formed.

Smectite clays tend to be associated with the subhumid to arid climatic regions that have produced grasslands in the United States. When found in the more humid regions, they are generally in soil formed from shale or in the residue from basic rocks.

Hydrous Mica Group. **Hydrous mica** (Fig. 5.5) has a slight structural difference from the primary mineral (mica) that is found in granite. Hydrous mica is probably derived by weathering of mica. It is associated with regions where weathering has not been severe and where the soil is neither very acid nor very basic. A member of this group is called **illite** after a location in Illinois where it was first identified. Hydrous mica is like montmorillonite in that it has a 2:1 lattice structure, but the lattice layers are held together by a mutual bond with potassium ions between them. This bonding minimizes the swelling and shrinking and results in good bearing strength for this clay and in reduced stickiness when wet. Illite has a lower capacity than does montmorillonite to hold plant nutrients.

The presence of hydrous mica in a soil does not make the soil unstable in the way that montmorillonite does. A predominance of hydrous mica clay in a soil indicates a lack

5.5 Layer lattice crystals of hydrous mica clays have a lower capacity to hold plant nutrients and to absorb water.

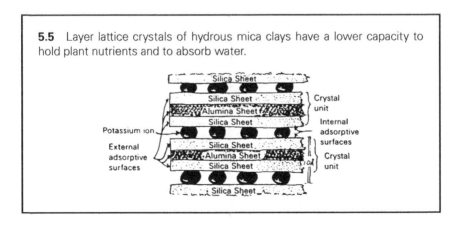

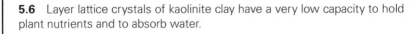

5.6 Layer lattice crystals of kaolinite clay have a very low capacity to hold plant nutrients and to absorb water.

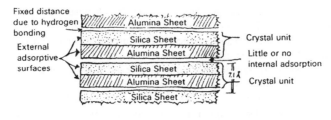

of severe weathering. Such clays are likely to be found in the cool climatic zones where precipitation is high enough to remove soluble salts from the soil.

When the interlayer potassium is completely removed by weathering, an expanding lattice 2:1 clay called vermiculite is formed. It does not shrink and swell as much as montmorillonite does. In vermiculite the negative charge is derived from the isomorphous substitution of Al^{3+} for Si^{4+} in the outer layer. As a result, vermiculite has a higher negative charge than does montmorillonite.

Kaolinite Group. The lattice of **kaolinite** clays is a 1:1 type made up of one silica and one alumina layer (Fig. 5.6). It can be seen that kaolinite has the least silica of any of the silicate clays. This is the result of the intense weathering that is characteristic of warm regions of the world.

One important property of kaolinite is the fixed spacing between the lattice layers. This is due to the attraction of hydrogen of the hydroxyl ions in an alumina layer for the oxygen in the adjacent silica layer. The bond between these lattice layers is called a hydrogen bond, and it is of great importance because it renders kaolinite less sticky and gives the soil a greater bearing strength than with other types of silicate clays. Kaolinite has a very low capacity to hold plant nutrients, and it absorbs less water than 2:1 clays.

Kaolinite, a favorite clay among potters, is most abundant in tropical and subtropical regions. Nearly pure deposits of kaolinite are valuable as sources for industrial materials. Large amounts are mined for use in the manufacture of bathroom fixtures.

Ammonium Fixation by Clays

Some 2:1 layer silicate clay minerals such as vermiculite and illite trap NH_4^+ and K^+ ions between interlayer spaces, resulting in a temporary immobilization of fertilizer N applied in a soil. Fixation occurs because the size of cavity left by oxygen in clays is sufficiently large to hold potassium and ammonium ions, but too small for other ions.

Identification of Layer Silicate Clay Minerals

Different clay minerals have contrasting properties; hence, it is important to identify them so that a soil's capabilities and limitations can be accurately predicted. The most common laboratory instrument used to identify silicate clays is the X-ray machine (Fig. 5.7).

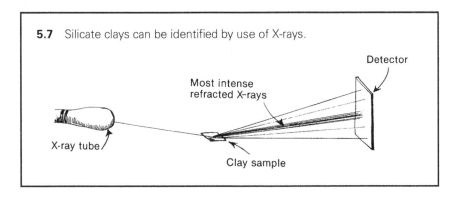

5.7 Silicate clays can be identified by use of X-rays.

Detector

Most intense
refracted X-rays

X-ray tube

Clay sample

Noncrystalline (Amorphous) Silicate Clays

When volcanic ash weathers in a relatively short time, some nearly **amorphous** (without structure) silicate clays form. Two of the clay minerals are allophane, which is spherical, and imogolite, which is threadlike. Except for their small size, they share few of the properties of the layer silicate clays listed previously. Their presence is germane to the classification of soils of volcanic origin.

Oxide Clays

To this point, consideration has been given to only silicate clays, but oxide and hydrated oxide clay minerals are also present in soils (Fig. 5.8). Normally, these are oxides of iron and aluminum, are amorphous, and are found most abundantly in soils formed from parent materials rich in iron and aluminum in tropical and subtropical regions where weathering has removed much of the silica from the clay fraction. Oxide clays have little or no crystallinity and very low capacity to hold plant nutrients. If iron oxides are not

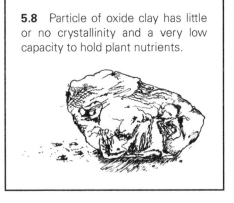

5.8 Particle of oxide clay has little or no crystallinity and a very low capacity to hold plant nutrients.

very hydrated, they give the soil a deep red color. Ferrihydrite is one of the iron oxide clays that is very hydrated; it is important in the classification of some volcanic region soils.

Cation Exchange

The characteristic of clay and humus to attract, hold, and release cations is called **cation exchange**.

Although most soil colloids have a net negative charge, no electrical charge in the soil goes unbalanced for very long. Electrical neutrality is maintained.

A soil colloidal system (primarily very fine clay and humus particles) has a double layer of charges. The inner layer is very closely associated with the surface of the colloidal particle discussed previously. The outer layer is formed by cations in the soil solution, which are attracted to the colloidal surfaces in proportion to the negative charges available. This means that a divalent cation such as calcium (Ca^{2+}) or magnesium (Mg^{2+}) can neutralize two negative charges of the colloidal particle, whereas monovalent ions such as potassium (K^+), sodium (Na^+), or hydrogen (H^+) can neutralize one negative charge each (Fig. 5.9).

In acid soils, aluminum ions (Al^{3+}), which may be combined with one or two hydroxyls (OH^-), can be attracted to colloidal surfaces. There may be many other cations attracted to the colloids in small amounts. Some of these are trace elements that are of great significance to growing plants.

Cations in the outer layer are sometimes called "swarm ions" because they resemble a swarm of bees around a hive, with the greatest concentration of bees close to the hive (Fig. 5.10). In a soil colloidal system, these cations become hydrated so their effective radius includes the water molecules.

The force of attraction of cations to the negatively charged colloidal surface differs among cations, depending on the number of positive charges and the effective diameters of the cations. The greater the number of positive charges (valence), the greater the force of attraction. For instance, Ca^{2+} has two positive charges and is attracted closer to the colloid than Na^+, which has one positive charge and tends to migrate farther from the surface of

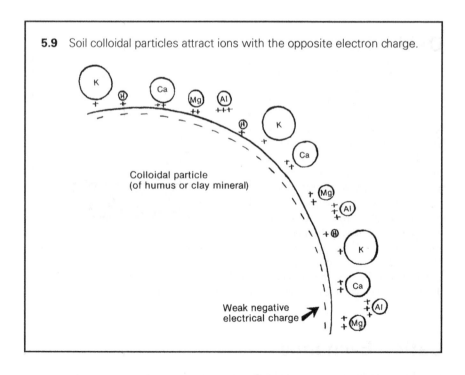

5.9   Soil colloidal particles attract ions with the opposite electron charge.

Colloidal particle
(of humus or clay mineral)

Weak negative
electrical charge

5.10 A "swarm" of positively charged ions around a negatively charged soil particle resembles bees around a hive.

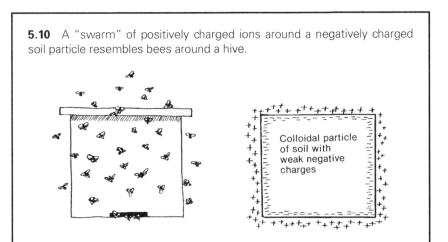

Colloidal particle of soil with weak negative charges

the colloid. The force of attraction increases in the order: $Na^+ < Mg^{2+} = Ca^{2+} < Al^{3+}$. For cations with the same number of charges, the force of attraction is dictated by the effective diameter of the cation. Cations with smaller effective diameters are attracted more strongly and they are held closer to the colloid than those with larger effective diameters. The force of attraction increases in the order: $Li^+ < Na^+ < K^+$. All the attracted cations are in constant motion, but attraction holds them tightly enough so they are not readily lost to water that is moving through the soil. They are adsorbed ions because they are held to the surfaces of the colloids. This action is very important to plant life because it keeps many nutrients within the root zone of the crops. Addition of cations to the soil, through acidification, liming, or fertilization, enhances the release of adsorbed cations into the soil solution as the new cations swap places on the exchange sites.

When there is sufficient moisture in the soil, cations in the soil solution can be lost from the soil profile by leaching. The nutrients that are easily leached are those that are less strongly held by soil particles. For instance, nitrate (an anion) will leach much more readily than calcium (a cation). In addition, monovalent cations, such as potassium, will leach more readily than divalent cations, such as calcium, since divalent cations are held closer to the exchange sites than monovalent cations.

In summary, cations are electrostatically attracted to the surface of the negatively charged colloids, but will diffuse away from the surface and toward the bulk soil solution based on concentration gradients. Anions, on the other hand, will be electrostatically repelled from the negatively charged colloid surface, but will diffuse toward the surface and away from the bulk solution based on concentration gradients. The exact nature of these processes is beyond the scope of this book, but they are very important in explaining many important soil properties.

Cations (e.g., Ca^{2+}) in a mineral fragment are released by weathering into the soil solution where they are attracted to particles of clay, around which they "swarm." By exchange with hydrogen ions coming from around roots, the nutrient ions finally reach the roots. There is an area called the oscillation zone in which ions are moving around roots and

5.11 A calcium ion (Ca²⁺) (left) migrates in solution toward a negatively charged soil particle to which two potassium ions (K⁺) have been previously attracted. The Ca²⁺ ion (right) changes places with the two K⁺ ions, which move on into the soil solution. An instance of cation exchange has occurred.

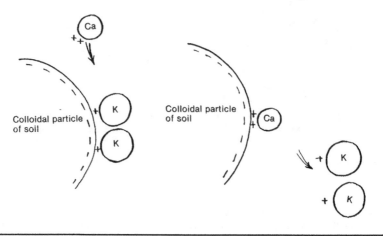

5.12 Cations move from a mineral, into solution, to the colloid surface, and on into the rootlet by ion exchange.

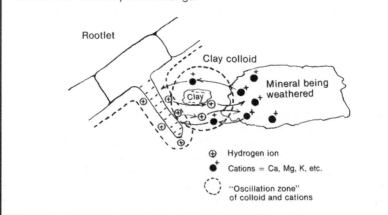

clay particles. This is the place of exchange, where one cation is replaced by another with an equivalent amount of charge (Fig. 5.11). For example, one divalent ion (such as Ca^{2+}) may replace another divalent ion (such as Mg^{2+}), or it may replace two monovalent ions (K^+ and K^+). When a plant takes cations from the soil solution (Fig. 5.12), it releases hydrogen ions (H^+) in exchange. For example, when one calcium ion is taken into the plant, two hydrogen ions are given off into the soil solution. Thus, electrical neutrality is maintained.

Cation Exchange Capacity

Cation exchange capacity (CEC) is defined as the sum total of exchangeable cations that a soil can adsorb. To quantify the negative charges on the soil colloids and therefore also the amount of cations attracted to those charges, it is essential to express the amount in standard units. The units are centimoles of charge per kilogram of soil material ($cmol_c/kg$). The "c" subscript before the slash indicates "charge." The quantities determined are designated as the **cation exchange capacity** (**CEC**). Typically, this measurement is determined on soil samples, but it may be made on other earthy deposits such as lake bottom sediments.

There are many variations in the laboratory determination of CEC, but the basic principles behind the methods are similar: (1) A known weight of soil is placed in a beaker and reacted with a solution containing only one type of cation, such as ammonium (NH_4^+). (2) When it has been established that all the negative sites on the colloids are satisfied with ammonium ions, the ammonium ions are replaced with another ion, and the ammonium ions replaced are measured. (3) The $cmol_c/kg$ of NH_4^+ determined represents the CEC of the soil sample.

Frequently the kinds of cations held on the colloidal system need to be determined. This can be done in a similar manner wherein the exchangeable cations in the soil sample are replaced with another cation and those removed are analyzed individually. The kinds of cations found on the colloidal system of most soils are quite predictable. They are calcium (Ca^{2+}), magnesium (Mg^{2+}), potassium (K^+), sodium (Na^+), and hydrogen (H^+). In some soils, aluminum (Al^{3+}) may also be very significant. When nitrogen is added to the soil in the form of ammonia (NH_3) or the ammonium ion (NH_4^+), the adsorption of NH_4^+ becomes important. The ranges in the CEC for pure samples of the clays discussed in this chapter are shown in Table 5.1.

For most agricultural soils, the CEC ranges between 3 and 20 $cmol_c/kg$. The very sandy soils are at the low end of the scale, and very clayey soils or organic soils may have a CEC much higher than 20 $cmol_c/kg$. Soil CEC is influenced by the amount of clay in the soil (texture), the type of clay present (mineralogy), the amount of organic matter present, and the soil pH. The CEC of the soil gives a strong indication of the ability of a soil to retain and release nutrients, but it does not replace a soil test for plant nutrients that are discussed later.

Table 5.1 The range in cation exchange capacity of some common clay minerals.

Type of clay	CEC in $cmol_c/kg$
Kaolinite	3–15
Illite	10–40
Montmorillonite	80–100
Vermiculite	100–150

A high CEC signifies a greater capacity to retain cations such as K^+, Ca^{2+}, Mg^{2+}, and NH_4^+ among others. This means that more nutrients are held on the soil colloids and the concentration of nutrients in the soil solution is low. This implies that while there are plenty of nutrients in the soil, the plants may not be able to take advantage of them, especially under conditions of low soil moisture, since nutrients held on the soil colloids have decreased mobility. These cations are also less likely to leach. A low CEC implies that fewer nutrients can be held by the soil and indicates a need for more frequent nutrient applications.

Humus as a Colloidal Substance

Humus is also part of the soil's colloidal system, and it releases valuable plant nutrients as it decomposes. Like clay, these microscopic particles carry negative charges to which cations are attracted. Humus has various chemical groups that can undergo loss of hydrogen at high soil pH to generate exchange sites for cation exchange. For this reason, the negative charge of colloidal humus particles can develop in several ways. One is from the migration of hydrogen ions (H^+) away from carboxyl groups that consist of carbon (C), oxygen (O), and hydroxyl (OH^-). They exist along the sides of the humus colloids (Fig. 5.13). The abundance of hydrogen ions in soils with a low pH restricts the migration of H^+ from the surface of humus colloids and reduces their CEC. For this reason humus has a pH-dependent charge rather than a permanent charge.

The CEC of humus ($\sim$300 cmol$_c$/kg) is at least twice or more times as high as that of silicate clays, so its value to the soil for crop production is enormous. Further, organic matter additions to sandy Coastal Plain soils whose clay fraction is dominated by kaolinite or hydrous oxides can greatly increase the CEC of these soils. This higher CEC may have an impact on how soil-test results for K, Mg, Ca, and other cations are interpreted. For instance, many state soil testing programs recognize this effect by separating state soils into groups for interpretation based on their CEC.

Anion Exchange

Up to this point, only those colloids that have a net negative charge have been discussed. The negative charge is the dominant condition. However, the word *net* in net negative charge implies that positively charged sites may also be present, though in lesser numbers, and this is indeed the case. In fact, certain acid tropical soils contain colloids with a net positive charge, and these colloids attract and exchange soluble anions just as negatively charged colloids attract and exchange soluble cations. Thus, these soils exhibit anion exchange capacity (AEC) instead of CEC. Soluble anions such as nitrate (NO_3^-),

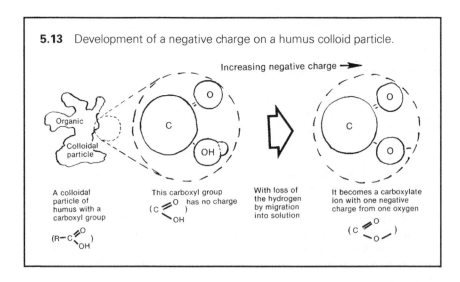

5.13 Development of a negative charge on a humus colloid particle.

chloride (Cl^-), and sulfate (SO_4^-) are held and exchanged on the positively charged surfaces, whereas the cations are repelled and remain in solution. Phosphate ($H_2PO_4^-$) is also attracted to these surfaces but is held much more tightly on surfaces of iron-, aluminum-, and calcium-bearing minerals by a specific adsorption mechanism that operates in either positively or negatively charged soils.

Surface charge becomes more positive (or less negative) as the soil acidity increases. Soils that show positive surface charges, therefore, are characteristically acid, and their colloid component is high in kaolinite, iron and aluminum oxides, and hydroxides but low in humus and expanding-layer silicate (smectite) clay.

In general, weathering removes the 2:1 clays and leaves 1:1 clays, such as kaolinite as well as iron and aluminum oxides. Under these conditions, the very acidic soils can develop positive charges. They are usually undesirable for crop production, not so much because of their charge, but because they often contain enough active aluminum and/or manganese to be toxic to plants. Many of these soils also have a very high capacity for fixing applied phosphorus in a form that is relatively unavailable to plants. Soils high in humus and smectite clay never go positive, but some tropical soils low in these components can be positive at pH 6 and below.

The usual problems with aluminum and manganese toxicity and phosphorus fixation in these soils make them generally undesirable. Also, these soils are so highly weathered that very few nutrients are released by further weathering. The AEC of most agricultural soils is small compared to their CEC. Therefore, anions such as NO_3^-, SO_4^-, and Cl^- are repelled by the net negative charge on soil colloids and they remain mobile in the soil solution. They are susceptible to leaching.

Soil Reaction (pH)

Soil reaction refers to the concentration of hydrogen ions (H^+) and hydroxyl ions (OH^-) in the soil solution, which are expressed in moles per liter. The term pH is a measure of the concentration and activity of hydrogen ions (H^+) in a system. It is defined as the negative log of the hydrogen ion concentration, or $-\log ([H^+]) = \log 1/H^+$. The more H^+ there is, the lower the pH and the greater the acidity. A pH $7.0 = -\log 0.0000001 = \log 1/10^{-7}$, pH $8.0 = -\log 0.00000001 = \log 1/10^{-8}$. The pH scale is the logarithm to the base 10 of the reciprocal of the hydrogen ion concentration. Thus, a unit decrease in pH results in a 10-fold increase in hydrogen ions, and a comparable decrease in hydroxyl ions. Thus, a pH of 6.0 is 10 times as acidic as a pH of 7.0 and 100 times as acidic as a pH of 8.0 (see Fig. 5.14).

Soil acidity can be in the soil solution (active acidity) and it can also be associated with the solid phases (reserve acidity) of the soil (iron and aluminum oxides, CEC, etc.). Reserve acidity is the main source of active acidity. Active acidity determines whether soil pH needs to be raised while reserve acidity determines how much lime must be applied to raise it. Soils with high clay and/or organic matter contents generally require more lime for a similar pH change as they tend to have higher reserve acidity than sandy soils with low organic matter.

The pH scale extends from 1 to 14, with pH 7 being precisely neutral. This means that at pH 7, the concentration of hydrogen and hydroxyl ions is equal. As hydrogen ions increase in concentration and hydroxyl ions decrease, the pH drops below 7 and vice versa. The pH ranges that might be encountered under natural soil conditions are illustrated in Figure 5.15. Soils are considered acidic below a pH of 5 and very acidic below a pH of 4.

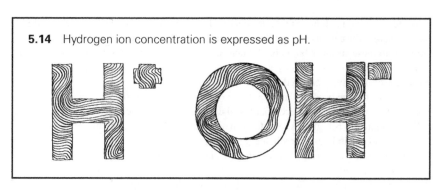

5.14   Hydrogen ion concentration is expressed as pH.

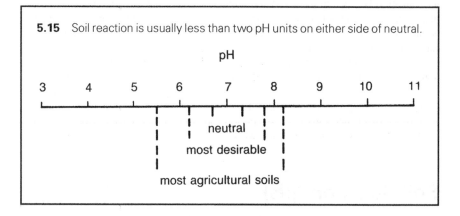

5.15   Soil reaction is usually less than two pH units on either side of neutral.

pH

neutral

most desirable

most agricultural soils

Conversely, soils are considered alkaline above a pH of 7.5 and very alkaline above a pH of 8. It is rare to find soils close to either of the extreme ends of the pH scale unless they have been contaminated by human activity.

Soil pH is an important chemical property that is most indicative of the general chemical status of a soil. For this reason, soil pH is the most commonly measured soil chemical property by farmers and urban homeowners alike. Its measurement can help in determination of: the amount of lime or sulfur to add to the soil, which fertilizers to use (some fertilizers will acidify the soil and others may raise the pH), and whether a particular pesticide can be tank-mixed with a particular fertilizer.

As discussed in Chapter 8, the use of manure and ammonium based nitrogen fertilizers, such as anhydrous ammonia, ammonium sulfate and urea, will gradually lower the soil pH. For this reason, it is useful to compare current soil test pH values to previous values to determine if there is a trend of soil pH change. Regular (every 2–3 years) monitoring of pH values in a field can help one consider if action is needed to remediate the pH.

In the laboratory, soil pH values are typically measured by mixing 10 g of air-dry soil with 20 ml of double-distilled water or 20 ml of 0.01 M $CaCl_2$ solution, and measuring the pH using an appropriate electrode connected to a pH meter. Soil pH measurement is a regular part of most, if not all soil test protocols.

The growth of crops is influenced by soil pH. Most field crops such as corn, small grains, cotton, and pasture grasses will grow satisfactorily over a relatively wide pH range—from 5.5 to 8.3—but the preferred range for best production is from pH 6.5 to 7.8. At this pH range, most plant nutrients are optimally available to plants. Furthermore, this range of pH is generally very compatible to plant root growth. Cranberries and blueberries, however, grow best in acid soils with pH 4.0–5.0, while alfalfa and sweet clover require a pH of 6.5 or above for best growth.

Plants are usually not directly affected by soils that range in pH below 5.0 or above 9.0; however, the indirect effect may be drastic. It is somewhat like having a fever that does not harm the body unless it goes to an extreme, but it certainly indicates that there is some sort of an infection that is at least temporarily harmful. Such soils are likely to produce poor crops for one or more of several possible reasons such as:

1. A lack of one or more plant nutrients.

2. Presence of plant nutrients in forms unavailable to plants. Low pH reduces the availability of phosphorus, calcium, magnesium, and Mo. High pH reduces the availability phosphorus and most micronutrients (Fe, Cu, Mn, Zn).

3. Diminished activity of beneficial soil microbes. For instance, the optimum soil pH for nitrification (the biological transformation of ammonium (NH_4^+) to nitrate (NO_3^-) is 6.6–8.0, and is markedly reduced at pH below 6.0. *Rhizobia*, the bacteria responsible for nitrogen (N) fixation in legumes, do not nodulate and fix N effectively under pH values less than 5.5. Since soil pH influences the activity of pathogenic microbes as well, farmers can adjust soil pH to manage some plant diseases. In general, the most abundant and diverse populations of soil organisms are found in near-neutral pH.

4. Abundance of ions toxic to plants. Low pH greatly increases the solubility of Al, Mn, Zn, and Fe, which are toxic to plants in excess. A high concentration of soluble Al also affects plant growth negatively through inhibition of Ca uptake and precipitation of P. In very acidic soils, the high concentration of H^+ ions causes irreversible damage to the uptake mechanisms of plant roots. Molybdenum is available at high pH and can be toxic to plants.

An unfavorable soil reaction can be remediated (see Chapter 9). The pH of alkaline soils can be lowered by adding elemental sulfur or sulfur-containing materials to alkaline soils. Acidic soils may be neutralized over time with the application of liming materials such as agricultural lime ($CaCO_3$) or dolomitic limestone ($CaMg(CO_3)_2$) among others.

When lime is applied to the soil, it dissolves to form carbonic acid and calcium hydroxide. Since carbonic acid is weak/unstable, it dissociates to carbon dioxide gas and water. The remaining calcium hydroxide dissociates into calcium and hydroxide ions. The calcium ions replace two hydrogen ions on the exchange sites, while the hydroxide ion reacts with the displaced hydrogen ions to form water, thereby resulting in an increase in soil pH.

The amount of lime required to adjust the pH of a soil to a desired value is termed "lime requirement." This amount is dependent upon the buffering capacity of the soil and how much the pH needs to be adjusted. The higher the amount of clay and organic matter, the higher the buffering capacity and more lime needed.

Remediation of soil pH has several benefits including prevention of the toxic effects of aluminum and micronutrients, provision of nutrients such as Ca and Mg, increased availability of essential nutrients, improvement in soil conditions for microbes, and improved soil structure.

Base Saturation

Base saturation refers to the percentage of base-forming ions (Ca^{2+}, Mg^{2+}, K^+, Na^+) that occupy the colloidal surfaces, or the CEC. Most cations in the soil are associated with the cation exchange complex of colloidal clay and humus discussed earlier. Under acidic conditions, comparatively few basic ions (calcium, magnesium, potassium) are present on the colloidal system, but there are many hydrogen ions. Aluminum ions on the colloidal system also give an acid reaction. For instance, in highly weathered soils of the Southeastern USA and the tropics, high rainfall in these areas leaches monovalent and divalent cations, leaving an abundance of Al^{3+} ions on the soil colloid. This results in low base saturation. In contrast, moderately weathered soils that formed from basic igneous rocks, such as the basalts, have relatively high base saturation.

In a laboratory, when the naturally occurring cations in a soil sample are replaced with another kind, the original cations can be collected and identified. The amount of basic exchangeable cations found may then be compared to the total amount of exchangeable cations. This is the means for calculating the percent base saturation. For example, on the colloidal clay particle in Figure 5.16 there are 25 cations, 15 of which are basic. Therefore, the percent base saturation is (15/25) × 100 = 60%. A high percentage base saturation is usually desirable for crops.

Keep in mind that numbers given here and in Figure 5.16 are for illustration only. The actual numbers are beyond comprehension. One mole of charge corresponds to Avogadro's number (6.02×10^{23}) so, for example, 1 $cmol_c$/kg of H^+ would be 1/100 of Avogadro's number (6.02×10^{21}). Figure 5.17 gives a comparison of this quantity to a spoonful of soil.

Buffering Capacity

In chemistry, buffering refers to a resistance to change in pH; consequently, a soil that is well buffered is one whose pH is not easily altered significantly. The cation exchange capacity of soils gives them most of their buffering capacity, as the vast majority of the basic and acidic ions are held to the surface of the clay and humus colloids are exchangeable. Thus, when basic ions are dissolved in the solution of a colloid-rich soil, there is very little increase in pH because hydrogen ions are released from the colloidal surfaces to neutralize the base that was added, that is the hydrogen and hydroxide ions react to form water (H^+ + $OH^- \rightarrow H_2O$). Well-buffered soils need more lime to raise their pH than those that lack

5.16 A colloidal clay particle has exchangeable cations around it. Each • (acid) or □ (base) represents billions of ions.

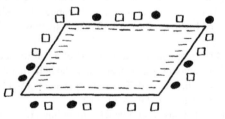

5.17 A spoonful of soil weighing 10 g (dry) contains about 1.2 quintillion (1.2 × 10²¹ exchange sites to which plant nutrients (Ca, K, etc.) can be held available for plant roots.

an abundance of colloids. The situation is analogous to the strength of an army—it is not so much related to the number of troops it has on the front line as it is to the forces it holds in reserve.

Reasons for Basic or Acidic Soil

Native soil pH is influenced by soil minerals and amount of precipitation. In arid and semiarid regions of the world, most soils are basic (alkaline) or nearly neutral for two reasons: The ions derived from weathering of minerals are predominantly base-forming ions, and there is not enough precipitation to leach them from the soil.

In humid regions of the world, leaching by precipitation causes the bases to be translocated deeper into the soil, and ultimately they return to the sea (Fig. 5.18). The effect of this process over the long span of geologic time is evident in deposits of limestone and other basic sedimentary rocks laid down on the sea bottom. The limestone deposits, which may be several hundred feet thick, have resulted from the concentration of calcium

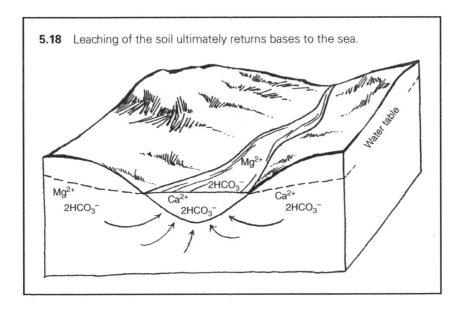

5.18 Leaching of the soil ultimately returns bases to the sea.

and magnesium carbonates by living organisms or by chemical precipitation in ancient seas. Other basic ions have been concentrated as salt beds when seas dried up. In many places, basic sedimentary deposits are covered with thick sandstone formations that may be somewhat acid or neutral in reaction. Other natural processes that contribute to soil acidification include the release of H^+ ions by plant roots, nitrogen fixation, and acid rainfall.

Although soil acidification is a natural process, some agricultural practices promote the acidification of soils. These practices include harvesting of crops, which removes cations such as Ca and Mg, the addition of manures, composts, and other organic residues, which release hydrogen ions during mineralization, and the generation of acid (H^+) from the nitrification of urea and ammonium-based fertilizers. A simplified equation showing how the conversion of ammonium to nitrate (nitrification) increases soil acidity is shown here:

$$2NH_4^+ + 4O_2 \rightarrow 2NO_3^- + 4H^+ + H_2O + Energy$$

Agricultural practices, such as tillage, also enhance soil acidification because they accelerate mineralization of organic matter as well as leaching. These effects, however, are not serious if steps are taken to counteract them by the addition of lime.

Soil Aggregation

The colloidal system of the soil not only is the center of chemical reactions but also has much to do with the physical structure of the soil. Most of the common soil cations (particularly Ca^{2+} and Mg^{2+}) attracted to colloids cause them to cluster into what is called a flocculated condition. The opposite of flocculation is dispersion, which means that the soil particles do not tend to cluster together when wet. Sodium ions (Na^+ disperse soil very effectively and cause the soil to flow together so that it becomes almost impermeable to water (Fig. 5.19). The reason is that the adsorbed Na^+ migrates far enough away from the colloids to leave the negative charges of the colloids unsaturated. Like charges repel, so the negative ($-$) particles disperse. The soil loses stability as a result. Entrance of sodium-rich water into earthen dams may cause them to fail.

The dispersed soil condition caused by sodium is very adverse to crop production. It is most commonly associated with slight depressions in otherwise level grasslands of semiarid regions as well as with cropland irrigated with water high in sodium. Sodium also

5.19   Soil is well aggregated by action of colloids rich in calcium ions (left). Soil runs together in a dense mass by action of colloids containing abundant sodium ions (right).

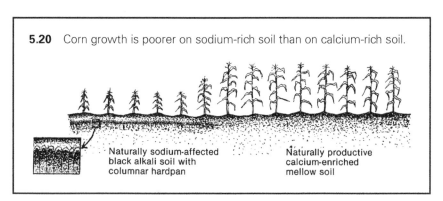

5.20 Corn growth is poorer on sodium-rich soil than on calcium-rich soil.

Naturally sodium-affected
black alkali soil with
columnar hardpan

Naturally productive
calcium-enriched
mellow soil

brings about a strong alkali condition that can dissolve humus, producing a dark crust on the soil surface when the water evaporates. Farmers call these "black alkali spots" and they are relatively unproductive (Fig. 5.20).

CHAPTER **6**

Soil Water

Many ancient civilizations left evidence that it was well understood how vital water was to the survival of their cultures. Nomadic tribes followed the seasonal rainfall patterns that affected the growth of forages for their grazing animals and of edible plants for their own consumption. Some of the earliest public works projects involved drainage and irrigation of lands to enhance crop production. The eventual collapse of some of these ancient civilizations has been attributed to poor management of water resources. Human reliance on a sufficient and timely supply of water for food and fiber production is no less critical today.

Water stored in the soil does several things. First, it is essential to plant growth. Nutrients move within the soil solution and are absorbed (taken up) from it by plants through the roots (see Chapter 8). Second, it is essential to the microorganisms that live in the soil and decompose organic matter and recycle plant nutrients (see Chapter 4). Third, it is important in the weathering process and soil formation by accelerating the breakdown of rocks and minerals to form soil and release plant nutrients (see Chapter 2). Fourth, water also plays a role in moderation of soil temperatures (Chapter 7). Fifth, water serves as an active factor in soil formation by translocating fine particles downward and dissolved substances both downward and upward.

Water in the soil influences the timing of many farming operations, such as when to till, when to plant, and when and how to apply herbicides and/or fertilizers. Soil water influences the choice of crops to be grown. In areas where rainfall and soil water are sufficient, corn may be grown. In areas with less rainfall and/or more evaporation, there is less soil water available, and a cereal crop or grain sorghum (milo) is more likely to be selected by farmers. To effectively manage available water resources, it is important to understand the processes of water movement in soils and uptake by plants.

Hydrologic Cycle

Hydrology is the study of the movement of water on the earth. The hydrologic cycle (Fig. 6.1) is used to summarize all the processes involving water in the environment. When the hydrologic cycle is considered on a global scale, it is common to begin with evaporation of water from the oceans. Evaporation also occurs from the land, and a small amount of water vapor comes directly from snow and ice in alpine and Polar Regions

Soil Science Simplified, Sixth Edition. Neal S. Eash, Thomas J. Sauer, Deb O'Dell, and Evah Odoi.
© 2016 John Wiley & Sons, Inc. Published 2016 by John Wiley & Sons, Inc.

6.1 The hydrologic cycle describes the flow of water in the environment.

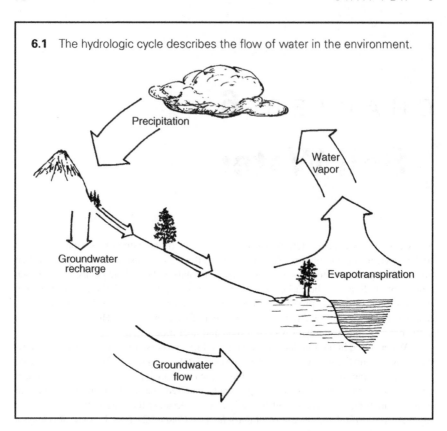

Precipitation

Water vapor

Groundwater recharge

Evapotranspiration

Groundwater flow

through **sublimation** (transformation of ice directly to water vapor). Water vapor in the atmosphere forms clouds and the water falls to earth in precipitation (rain, snow, sleet, and hail). Precipitation that falls on the ocean can be evaporated again. Snow that falls in polar or mountainous regions may be stored for decades or centuries before it melts.

Some of the precipitation that falls on land is intercepted by vegetation and evaporates back to the atmosphere but most of it reaches the soil surface.

Precipitation that reaches the soil can either enter the soil or run off the land to a surface water body (stream, marsh, or lake). Surface water eventually evaporates, seeps farther into the earth, or flows back to the oceans, where it can evaporate and start the cycle again. Water that enters the soil is of most importance to plant growth. This water can evaporate from the soil surface, be absorbed by plant roots to be utilized by the plant or transpired (evaporate from leaves), or pass through and out of the root zone to become part of the groundwater.

The global hydrologic cycle is very complex and involves processes that occur on large scales (precipitation) and over long periods of time (melting of glaciers). Nonetheless, parts of the hydrologic cycle have strong implications for food and fiber production. For instance, in several areas of the world, water from melting snow and ice is captured in reservoirs and used to irrigate crops sometimes hundreds of miles (kilometers) away.

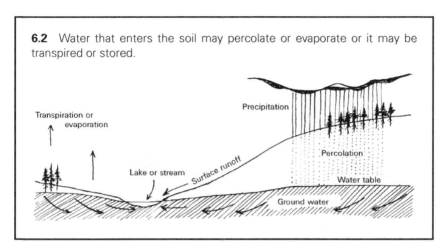

6.2 Water that enters the soil may percolate or evaporate or it may be transpired or stored.

Soil Water Budget

Although soil water is just one component of the hydrologic cycle, it represents the crucial reservoir of water for the growth of most plants. An easy way to monitor water in soils is to consider the soil's water budget. Just as a person may have a financial budget with inputs (income, investments, etc.) and outputs (food, clothing, shelter, etc.), the soil has a water budget. In agricultural settings there are two inputs: precipitation and irrigation. Water from precipitation and/or irrigation can either move into the soil (**infiltration**), or it can run across the soil surface to a stream, marsh, or lake (**runoff**).

What happens to water after it enters the soil? It can be stored in the root zone for future use by plants or move through the soil and out of the root zone (**percolation**) and eventually to groundwater. It may also evaporate, either directly from the soil or from plant leaves (**transpiration**) after being absorbed by roots. Thus, the soil's stored water is the difference between the sum of all inputs and the sum of all outputs. The water budget has two inputs (precipitation and irrigation) and three outputs (runoff, evaporation, and percolation) (Fig. 6.2).

The water budget can be put in equation form:

$$S = (P + I) - (R + E + D)$$

where S is the amount of stored water, P is precipitation, I is irrigation, R is runoff, E is evaporation and transpiration, and D is percolation. Each term in the water budget equation would have linear units of measurement such as inches (millimeters). Typically, water budget analyses are completed for months or years so managers can analyze trends in each term and consider options to optimize water use.

The amount of water that ends up in each term is partly determined by climate and partly by properties of the soil and the requirements of the plants growing in the soil. Of course, humans also have the opportunity to manage the movement of water by choosing which crops to plant, when and how much to irrigate, types of tillage and residue management practices to follow, or to provide drainage of excess water.

6.3 Soils with large pore spaces, such as sandy soils and well-granulated types, usually have high infiltration and percolation rates, whereas those that have small pore spaces or are in poor physical condition have low infiltration and percolation rates. Runoff occurs if the rate of rainfall exceeds the water infiltration rate.

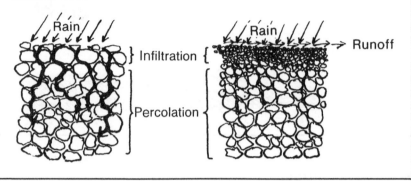

Infiltration and Runoff

Precipitation or irrigation that reaches the soil surface is partitioned between infiltration and runoff. The rate of infiltration varies with the texture and physical condition (structure and porosity) at the soil surface. Sandy soil, because of its relatively large pore size, has a higher infiltration rate than clay soil with its smaller pore size (Fig. 6.3). If the physical condition of the soil is poor, the infiltration rate is reduced. A sandy soil may have an infiltration rate greater than 1 in. (2.5 cm) per h, whereas some clayey soils require more than 12 h for 1 in. (2.5 cm) of water to infiltrate.

Figure 6.4 gives an indication of the rates of runoff and infiltration for a hypothetical rainstorm. The rate of infiltration needs to be known when designing a drainage or irrigation system. One way to estimate infiltration would be to observe how long it takes before water starts to run off (if it ever does) during a rainstorm. If the rate of water infiltration into soil is less than the rate at which rain falls, water accumulates on the soil surface. If enough water accumulates to fill the small depressions at the soil surface, runoff begins. If the amount of rainfall and the duration of the storm are known, then the infiltration rate can be estimated.

As runoff water flows downslope across the land surface, it gathers momentum and picks up soil particles, which results in soil erosion. It is generally desirable to hold as much of the rain as possible where it falls to provide water for crops and to protect the soil from erosion. On some soils in humid regions, however, it is necessary to encourage runoff through a surface drainage system that prevents the soil from becoming waterlogged. It is impossible to avoid all erosion, but it is important for it to be minimized. Soil erosion will be discussed further in Chapter 10.

A farmer can influence the infiltration rate of a soil by keeping a protective vegetative cover on the surface and by maintaining good soil structure, both of which help conserve water and soil. By keeping the soil in good physical condition, the topsoil becomes full of "crumbs," which are stable, spongy aggregates. With this condition of the soil, water moves

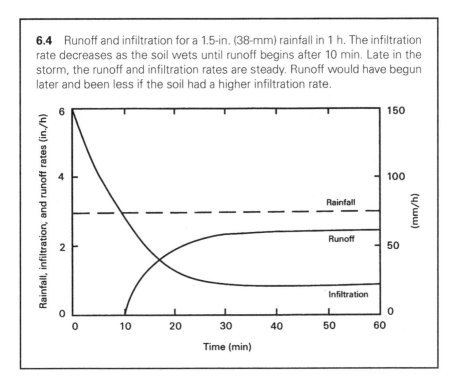

6.4 Runoff and infiltration for a 1.5-in. (38-mm) rainfall in 1 h. The infiltration rate decreases as the soil wets until runoff begins after 10 min. Late in the storm, the runoff and infiltration rates are steady. Runoff would have begun later and been less if the soil had a higher infiltration rate.

easily into and through the soil. Such aggregates form when adequate organic matter is present and excessive tillage is avoided. The farmer who depletes the soil of organic matter by removing crops without returning plant residues or manure is likely to decrease the infiltration rate. When raindrops strike exposed soil, especially without the spongy aggregates, the soil is beaten to a paste and a seal forms on the surface. The soil surface then tends to shed water like a roof.

When a sealed-off surface becomes dry, it forms a brittle crust that can inhibit emergence of seedlings. Seedlings such as those of beans and potatoes are strong enough to break through, lifting pieces of the crust like so many trapdoors opened from below. But seedlings of small-seed crops such as oats and even the larger seedlings such as corn may perish without ever emerging (Fig. 6.5). Sometimes farmers have to break this crust with light tillage after planting the crop, but this may be only partially successful.

Evaporation
Evaporation is the transformation of water from liquid to vapor regulated by solar energy, wind movement, and humidity. Soil water can evaporate directly from the soil or it can be absorbed by roots and evaporate from stomates on the leaves of plants. The process of evaporation from stomates is called **transpiration**. Evaporation from soil and transpiration by plants may be combined and called **evapotranspiration** (Fig. 6.6).

6.5 If a plant seedling is not strong enough to lift the soil crust, it dies.

Soil crust after a rainstorm

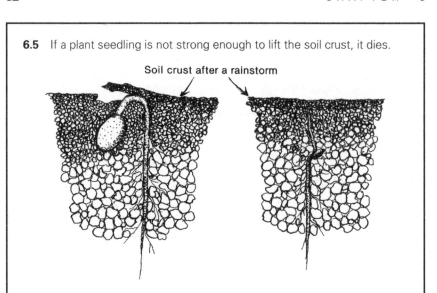

6.6 Soil water returns to the atmosphere by evaporation from the soil surface and by transpiration from plant leaves.

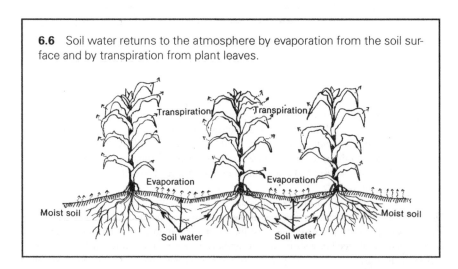

In many temperate and drier regions, at least half the water lost from farmland is by evaporation. Therefore, farming practices are often designed to reduce this loss and conserve moisture. One effective practice is to leave the plant residue from the previous crop on the soil surface. These crop residues reduce evaporation by shading the soil and blocking water vapor movement.

In small plots of high-value crops, mulches are often used to hold the soil water for plants and thus reduce evaporation (Fig. 6.7). Many kinds of mulches have been used: straw, corncobs, gravel, and plastic. All can be quite effective. The selection of one type of mulch over another depends on the specific use and availability of the material. Organic materials such as straw are preferable in situations where the mulch can be incorporated into the soil

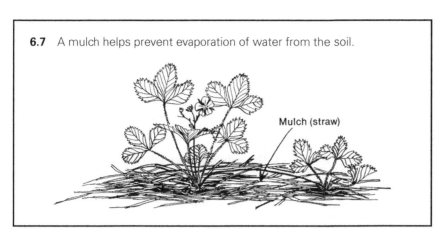

6.7 A mulch helps prevent evaporation of water from the soil.

Mulch (straw)

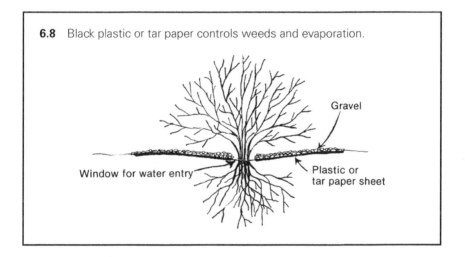

6.8 Black plastic or tar paper controls weeds and evaporation.

Gravel

Window for water entry

Plastic or
tar paper sheet

after each crop. Sand and gravel have the advantage of allowing a higher percentage of small rains to infiltrate into the soil rather than being absorbed by the mulch. This can be an advantage around fruit trees and ornamental plantings.

The use of black plastic as a mulch is popular in vegetable production because it effectively controls both weeds and evaporation. A variation in this practice is to form plastic or tar paper into a shallow cone around the base of a tree or shrub and cover it with a few inches of gravel. This allows rainwater to enter the soil near the trunk and leaves no place for weeds or grass to grow in hard-to-mow places (Fig. 6.8).

Farmers in dry areas where wheat is a leading crop utilize the principle of mulching by leaving much of the plant residue on the soil surface when tilling after harvest. Creating a dust mulch by frequent tillage of fallow land is now often discouraged because it has been found that little water is conserved by this practice and that soil may be left susceptible to severe wind erosion.

Percolation

In 1856, an engineer named Henry Darcy was the first to describe how water moves through a saturated soil (all pores filled with water). He developed his theory by observing the flow through a sand filter used to purify water in the French city of Dijon, famous for the production of mustard.

The force of gravity causes water to move downward through the soil, particularly in larger pores. Gravitational water percolates until it is adsorbed by drier soil below or it reaches the water table. The **water table** is the level in porous subsurface materials below which all pores are filled with water. This may be within the surface soil, in buried sediments, or in the deep bedrock. **Groundwater** is the water below the water table. The amount of water that percolates through the root zone to groundwater is referred to as **groundwater recharge**. The water table surface has a slight slope (much less undulating compared to land slopes) allowing the water to flow laterally below ground. Porous soil or rock layers that are saturated can be important sources of well water. Such layers are called **aquifers**.

As water moves downward through the soil to reach the water table and the ground water, it may encounter a soil layer that restricts downward movement causing water to build up and form a perched water table. If the perched water table is within the root zone, plant roots growing in this zone may be deprived of oxygen, thus impacting plant growth.

In a flat area, water flow in a uniform soil will be primarily downward. On steep slopes or when there is a gentle slope with a restricting or more conductive layer in the soil (like a layer of clay or gravel), water movement may still be downward but some water may also move downslope or laterally. Lateral water flow in mountainous areas may come to the surface again as discharge from flowing springs. In all cases, plant uptake draws water out of the soil, altering water movement in the soil near the roots.

Wetlands, lakes, and streams in humid regions are often places where the water table comes to the surface and groundwater discharge takes place. Wetlands in drier regions, however, are often groundwater recharge areas, where surface runoff collects and infiltration occurs. Thus, wetlands in these areas serve an important role in replenishing aquifers.

Soil Water Storage and Movement

Two forces impact soil water storage and movement. They are; (1) gravity and (2) attraction of soil particles for water. When soil is saturated with moisture, gravity is the dominant force in moving water in large pores deeper into the soil. At the same time that gravity is pulling water downward, the soil particles are attracting water in all directions by the forces of adhesion and cohesion. **Adhesion** is the attraction of a surface for water (e.g., the surface of the soil particle), and this force is quite strong. **Cohesion** is the attraction of one water molecule for another. The two forces combine so that water is held within small pores between soil particles. Because one of the forces is the attraction a soil particle surface has for water, it follows that a soil with very small particles and more surface area such as a clay attracts and holds more water than a soil with large particles and less surface area such as sand.

When a soil dries, adhesive forces begin to dominate, and the water remaining in films is held very close to the soil particles with greater force due to the surface attraction for water. Consequently, water movement is very slow. Water movement in unsaturated soils involves a complex combination of gravity, relative amount of water, and adhesive/cohesive

6.9 Water moves from soil particles with the thickest water films to soil with the thinnest. As the plant root absorbs moisture, water tends to move toward it (capillary movement). Plant roots also grow and extend into zones with more moisture.

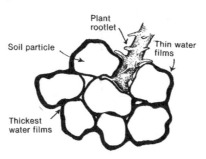

forces. It was not until 1907 that an American, Edgar Buckingham, was able to accurately describe water flow in unsaturated soils.

As a plant root absorbs water, it takes some from the film surrounding the adjacent soil particle. Due to cohesion and adhesion, water moves from particles with thicker films to particles with thinner films that are next to the roots. This is called **capillary movement** (Fig. 6.9). Capillary water may move to the roots from any direction—up, down, or laterally. An important fact is that water moves in the soil toward roots to provide plants with water. Some essential nutrients can also move with the water. Capillary movement, however, is normally very slow in soil, so plants must continually extend their roots into moist pores to absorb water. The soil thus has a critical role in agriculture as it acts as a storehouse of water for plants to use until the next rain or irrigation.

Water Use by Plants

Water is essential to all forms of life—both plants and animals. Some plants have low water requirements and are called xerophytes (*xero* means little or none and "phytes" from the word *phyto*, which means plant). Some have high water requirements and are called hydrophytes (*hydro* means water). Plants with moderate water needs are called mesophytes (*meso* means intermediate).

Plants need water to form certain compounds. For example, six parts of water are required for each simple sugar produced. The process of forming simple sugars, called photosynthesis, involves the splitting of water (H_2O) into hydrogen and oxygen. The hydrogen combines with carbon dioxide (CO_2) to form sugars, and the oxygen is discharged into the atmosphere through openings called stomates in the leaves of plants.

Much of the water stored in the soil is used by a plant. The amount of water stored in the soil and the amount available to plants vary with the texture and structure of the soil. When soil water content is near saturation and gravitational forces are dominant, most water in the soil is readily available to the plant.

The maximum amount of water in a soil held against the force of gravity is called the **field capacity**. As water is used by plants or evaporates, the water film around soil

6.10 The water films in A are thickest and the soil is nearly saturated; at B it is about at field capacity; and the thin films in C represent the wilting point.

(A)   (B)   (C)

6.11 Soil water between field capacity and the wilting point is available to the plant.

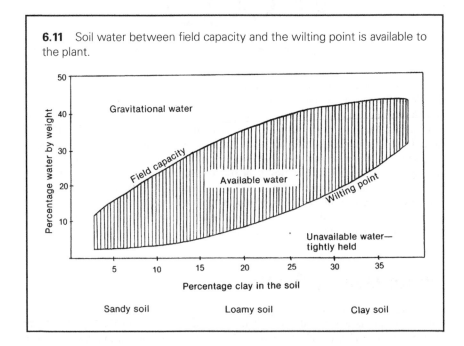

particles becomes thinner, is more tightly held by the particles, and is more difficult for the plant to absorb. Eventually, the attraction between the soil and the water is greater than the plant's ability to absorb it. This amount of water in a soil is called the **wilting point** because the plant can no longer absorb enough water to maintain transpiration and sustain life (Fig. 6.10).

Between these two points, the field capacity and the wilting point, water is available to the plant (Fig. 6.11). The amount that is available varies with soil texture (and physical condition). For example, a sandy soil (which has large particles and low surface area) may store about 1 in. of water per foot of soil depth (83 mm/m) and most of the water would be available to plants. A clay soil (which has small particles and high surface area) may hold 4 in. of water per foot of soil depth (333 mm/m), but because of the strong attraction

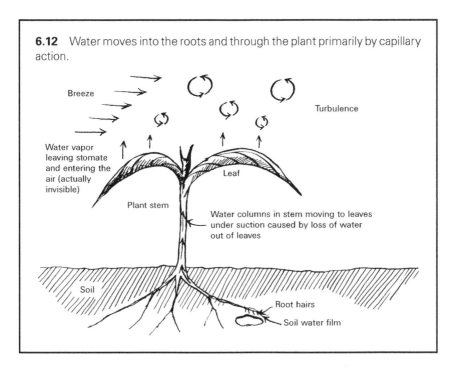

6.12   Water moves into the roots and through the plant primarily by capillary action.

Breeze

Turbulence

Water vapor leaving stomate and entering the air (actually invisible)

Leaf

Plant stem

Water columns in stem moving to leaves under suction caused by loss of water out of leaves

Soil

Root hairs

Soil water film

of clay particles for water, only 1 in. of these 4 in. may be available for plant use. Soils with the greatest amount of available water are usually those with a loamy texture and good structure.

Only a very small percentage of the water absorbed and utilized by a plant is for photosynthesis. Water's principal function is to transport nutrients and plant compounds in solution, either upward from the plant roots to the upper leaves or downward into lower leaves or the root system. Most of the water taken up by a plant eventually evaporates at the stomates (Fig. 6.12). Water lost from the plant due to evaporation from the stomates is called transpiration and, especially in hot weather, transpiration helps cool a plant. Less than 1% of the water absorbed by a plant is used in forming plant compounds; the rest is lost via transpiration.

A plant's water use efficiency is determined by measuring the amount of water required to produce a certain weight of dry plant tissue. It takes approximately 500 lb (225 kg) of water to produce 1 lb (0.45 kg) of wheat (foliage plus grain). Only 5 lb (2.25 kg), or 1% of this amount actually becomes part of the plant. Alfalfa uses more water, requiring about 850 lb (385 kg) of water per pound (0.45 kg) of dry matter; while grain sorghum, an efficient water user, may require less than 300 lb (135 kg) of water per pound (0.45 kg) of dry matter.

Drainage

It is a common occurrence in many regions of the world for the soil to contain too much water during rainy seasons of the year or during winter when evaporation is low. If

6.13 The water table can be lowered to the level of the subsurface drainage network.

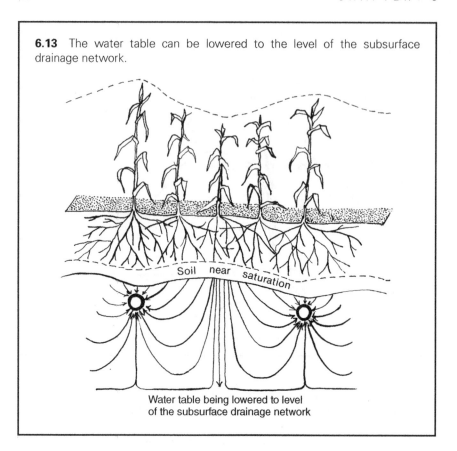

Water table being lowered to level
of the subsurface drainage network

the soil is waterlogged too long during the growing season, roots die from lack of oxygen or from accumulation of toxic compounds. To rid the soil of excess water, drainage systems have been installed on millions of acres (hectares) of land. Drainage systems can involve subsurface or surface practices or a combination of both. Remarkable increases in crop yields can occur when naturally wet soils are drained.

Subsurface drainage is the practice of burying a network of perforated pipes horizontally in the soil. The pipes intercept percolating water, or capture the water in a perched or true water table. The water entering the pipes moves laterally to a surface outlet such as a drainage ditch. Drain pipes or tiles were originally made of short sections of concrete or clay, but long lengths of flexible plastic tubing are now more popular. The tubing is installed at a depth of 2–6 ft. (0.6–1.8 m) and has a slight downward gradient to the surface outlet. Subsurface drainage functions only when the soil is saturated, so that water can flow from the large pores in the soil into the gaps between sections of clay or concrete tile or through holes in the plastic tubing and on along the tile or tubing to the outlet (Fig. 6.13).

In some cases, a vertical tube is installed from the subsurface drain to the soil surface to allow water ponded on the soil to enter the drain without percolating through the soil. These surface inlets or intakes are commonly found in areas with small depressions that fill with water during heavy rains or spring snowmelt.

6.14 Different types of subsurface drainage systems.

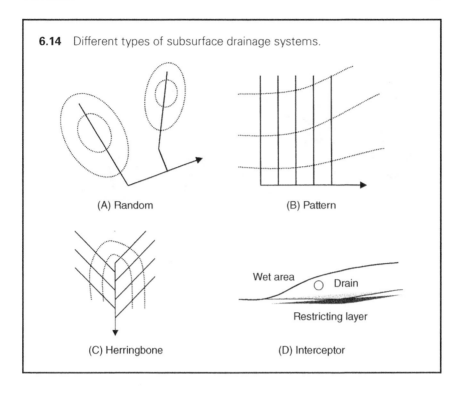

(A) Random   (B) Pattern

(C) Herringbone   (D) Interceptor

Wet area — ○ Drain

Restricting layer

Subsurface drainage can be installed in different patterns, depths, and spacing depending on land slope and location of the outlet (Fig. 6.14A–D). The random design (Fig. 6.14A) is used where there are isolated wet areas. Drain lines are run under each area with perhaps a surface inlet in the larger depressions. Pattern (Fig. 6.14B) and herringbone (Fig. 6.14C) drainage patterns involve uniform distances between multiple drain lines. The pattern used depends on slope of the land and desired depth of unsaturated soil. An interceptor drain (Fig. 6.14D) is used to intercept lateral flow down a slope that may be creating a wet spot.

Surface drainage involves digging channels in the soil and sometimes also shaping the land surface so water will run over the surface into the channels. Surface drainage is used on soils that have layers with low permeability or in very flat areas like the Red River Valley of North Dakota and Minnesota. In these areas, water either cannot move through the soil fast enough or the slope of the land is nearly flat for subsurface drains to flow effectively. Combination surface/subsurface drainage systems involve subsurface drains using surface drainage channels for their outlets.

To have an outlet for subsurface drains, some natural stream channels have been straightened and/or deepened (Fig. 6.15). This practice is called channelization, and is often criticized for impacting water quality and wildlife conservation. Draining wetlands that should be preserved for wildlife habitat or for water quality protection is often undesirable. In these cases, channelization is usually undesirable. However, on agricultural land that is in crops and pasture, channels can be beneficial to both agriculture and wildlife. When intermittent waterways are deepened, a permanent stream may be formed. As a result, fish thrive, and birds, mammals, and reptiles find an improved environment along the channel banks. Crops also flourish, and they too provide food and shelter for wildlife.

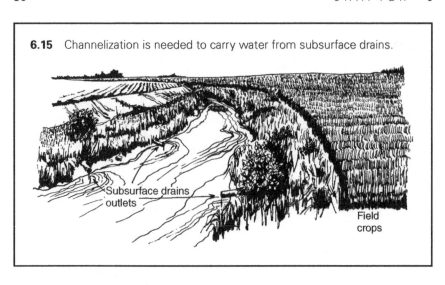

6.15   Channelization is needed to carry water from subsurface drains.

Subsurface drains
outlets

Field
crops

To help offset the loss of wetlands due to farmland drainage and channelization, artificial wetlands can be created. Shallow depressions are made in the soil in low-lying areas and wetland plant species are seeded or transplanted into the depression. Water from runoff or subsurface drains is directed into the wetland. The wetland may be designed to be flooded the entire year or only after spring snowmelt and large rainstorms. These constructed wetlands; (1) provide wildlife habitat, (2) may help reduce chemical and sediment transport to streams, and (3) increase groundwater recharge.

Irrigation

Since ancient times, various civilizations have utilized water from rivers and wells to ensure more reliable crop production. Today, about 17% of the World's cropland is irrigated. Because yields on irrigated land are usually high, irrigated cropland accounts for a disproportionately large part of crop production. In countries where rice is the staple diet, the people rely on irrigated agriculture for most of their food. Many of these countries have a monsoon climate where the annual rainfall is high but most rain falls during a few months and is followed by an extended period of limited rainfall. Even in the temperate regions, timely rains can be quite unpredictable from year to year so irrigation agriculture is common.

Different irrigation systems are used based on the crops being grown, the source and quality of water, and the level of technology available. Flood irrigation involves flooding the entire field with water from canals or pipes (see Fig. 6.16). Paddy rice production is an example of flood irrigation. A berm around the field contains the water and sometimes the plants are grown in rows on ridges with the water flowing between the rows. With sprinkler irrigation water is sprayed onto the crop from nozzles suspended above the canopy. Often the water source is a well and electricity or an engine is needed to pump the water and/or move the sprinklers. Sub-irrigation can be accomplished with small tubes buried in the soil through which water is pumped. Water seeps into the soil through emitters spaced along the tubing. Another type of sub-irrigation involves putting water back into subsurface drain

6.16 Examples of types of irrigation systems: (A) surface or flood, (B) sprinkler, (C) sub-irrigation, and (D) drip. Photos courtesy of USDA NRCS.

tubes to re-wet soil in the root zone. Drip irrigation utilizes small tubes on the soil surface with drippers or micro-sprinklers placed near the plants. Water is pumped through the tubes and the amount of water applied depends on the length of time and the number and size of the drippers or micro-sprinklers. Drip irrigation is most often used with high value crops and where there is a very limited availability of water for irrigation (see Fig. 6.16).

The first two conditions to consider for irrigation feasibility are a source of water and adequate soil drainage. The quality of irrigation water is important because it may have long-term impacts on soil properties. The two common qualities of irrigation water that are of concern are high total salt content and a high-concentration of sodium. In some cases, other ions such as boron, lithium, and selenium may be present at toxic levels in the proposed water source. Sometimes special water-utilization techniques can be employed to enable the use of water of marginal quality. In some instances, irrigation water may contain beneficial ions. Nitrate nitrogen (NO_3^-) might be present in irrigation water at high concentrations, such as in the Platte River valley of Nebraska. In these instances, a credit should be given for the amount of nitrogen applied during irrigation.

All irrigation water contains salt, and when irrigation water evaporates, the salts tend to accumulate and might negatively impact plant growth. This is when the internal drainage of the soil becomes important. If salts accumulate, the only way to remove them is by applying more water. When this water moves downward, the salts dissolve and move with the water. This is called **leaching**. Water used for leaching can be either from rainfall or irrigation. The removal of salt from irrigated soil may create other problems. If internal drainage of the soil is good, salts can be leached down through the root zone and either accumulate there or continue to move downward. If soil drainage is poor, salts can move off the soil surface as water moves across it (as in rice production). In some soils with poor internal drainage, subsurface drainage systems may be installed. As water moves through the soil into the drainage system, salts move with it (such as in the Imperial Valley of California). Regardless, the salt concentration in the water increases as it moves across or through the soil. The water coming out may contain three to four times the salt content compared to the water going into the field.

Discharging water with high salt content into the drainage system of an area can create problems for those downstream from the point of discharge. If salt concentration is too high, it can be detrimental to human and livestock consumption or for reuse as irrigation water for crop production.

If sodium concentration is high in the irrigation water, the soil will develop a high exchangeable sodium level. Sodium is attracted to a lesser extent to colloidal surfaces than is Ca^{2+}, but it takes only about 15% Na^+ on the exchange complex to cause a dispersal of clay particles. Soils with high exchangeable sodium will normally develop surface crusts and soil aggregates will disperse. These conditions decrease the rate at which water moves through the soil, which interferes with drainage and salinity control. Scientists at the U.S. Salinity Laboratory developed the **sodium adsorption ratio** (**SAR**) to characterize the sodium status of irrigation water and soil solutions:

$$SAR = [Na^+]/([Ca^{2+} + Mg^{2+}]/2)^{1/2}$$

where the concentrations of Na^+, Ca^{2+}, and Mg^{2+} are expressed as moles of charge/L. SAR values are not used alone but rather with other measures of salinity or sodicity. A SAR below 13–15 for a soil extract is generally considered acceptable.

About three-quarters of the irrigation water in the United States comes from surface waters such as rivers and reservoirs where rainwater and/or snowmelt have been impounded. The remaining quarter comes from underground aquifers. Some wells tap shallow aquifers that are replenished by precipitation annually, such as in the Central Sands of Wisconsin, but others "mine" deep water that was trapped in aquifers thousands of years ago and is not being replenished to an appreciable degree. This situation, referred to as **overdraft**, is occurring in parts of the High Plains of Texas, where the water table is dropping. This potential problem exists where the Ogallala aquifer, which reaches from South Dakota to Texas, has been supplying irrigation water since the 1930s. The solution for extending the life of the Ogallala aquifer, while maintaining agricultural production, is improved efficiency in the use of irrigation water.

Water Conservation

The recognition of high-quality water as a valuable resource has led to extensive research on improving irrigation efficiency. Flood irrigation has a lower water use efficiency as significant water can be lost to evaporation, seepage below the root zone, and as tailwater (water that flows out of the field). Sprinkler irrigation typically has a higher water use efficiency although there can still be large evaporation losses. Sub-irrigation and drip irrigation are the most efficient types of irrigation. With these systems the water is applied slowly where roots can reach it so very little water is wasted. However, the cost of these systems is high and their installation and operation require greater management skill.

Other, general water-conserving techniques now in common use include:

1. improved timing of water application based on measured soil moisture in the root zone (often continuously monitored with sensors buried in the soil),
2. plastic lining of supply ditches,
3. selection of crops and varieties with higher water use efficiency,
4. optimized plant population density, and
5. attention to plant nutrition and health maintenance.

To minimize water loss by evaporation, plant residues or mulch can be especially effective.

CHAPTER **7**

Soil Temperature

All plants need sunlight to grow. Light from the sun supplies the energy needed for photosynthesis and also warms the soil and air in which crops grow. Soil temperature affects almost every physical, chemical, and biological activity that occurs in the soil. Management decisions, such as when to plant, are often based on soil temperature. Knowledge of the flow of energy in the soil–plant–atmosphere system helps to understand how plants respond to climate conditions.

Importance of Soil Temperature

Plants

Soil temperature is a greater influencing factor in plant growth rates than above ground air temperatures. Soil temperature influences date of planting, time to germination, and number of days for a crop to mature. Ideal soil temperatures for germination vary depending on the crop and the seed characteristics. Most seeds (provided other conditions are ideal, such as adequate soil moisture) require minimum soil temperatures of 40–60°F (4.5–15.5°C) to germinate. In some cases extreme soil temperature may restrict germination and plant growth and extremely high temperature can cause severe heat stress to young seedlings. Cooler soil temperatures may diminish the ability of the roots to absorb water and nutrients. In normal situations, with increased temperatures, germination and growth of seedlings are enhanced, and root development is faster.

Optimum soil temperatures for plant growth are generally higher for plants that have evolved in warm climates than those evolved in cooler climates. Crops like cotton grow well in warm soil conditions while potatoes, rye, and oats prefer cooler soil conditions. Similarly, different trees have preferences for cooler or warmer soil conditions. In forested areas of the Midwestern United States, a dense oak–hickory–maple forest is found growing in the cooler soils of the north-facing slope, while south-facing slopes (warmer soils) often have a sparse stand of red cedar and burr oak with grass between the trees.

Soil Science Simplified, Sixth Edition. Neal S. Eash, Thomas J. Sauer, Deb O'Dell, and Evah Odoi.
© 2016 John Wiley & Sons, Inc. Published 2016 by John Wiley & Sons, Inc.

Microorganisms

Microorganisms are vital to the breakdown of plant and animal residues (organic matter) with the resultant release of essential plant nutrients including nitrogen, phosphorus, and sulfur. Soil temperature influences the growth and activity of microorganisms in the soil. Each organism has an optimum temperature at which its metabolic activity is at its highest level (most rapid breakdown of organic matter). Normal soil temperatures are generally cooler than the optimum for most organisms. As a result any increases in temperature (approaching the optimum temperature) could result in increased microbial activity and faster release of plant nutrients. While warmer soil temperatures can help the beneficial soil organisms, the same is true for the soil organisms responsible for plant diseases. Microbes are also responsible for breakdown of many of the organic wastes and pesticides. Warmer temperatures could help the microbes detoxify a soil faster. Extremes in soil temperature can have a slowing effect on microorganism and plant growth. Extremely high temperatures can destroy pathogenic organisms and weed seeds as is done in composting operations.

Solubility of Minerals

Soil temperature has an influence on the dissolving of minerals by water. Warm water hastens the dissolution of minerals, making nutrients from these minerals available for plant use. High soil temperature and adequate soil moisture mean, in most cases, more nutrients in solution. Soils in climates where rainfall is excessive and soil temperatures are high generally have low nutrient levels because the nutrients from dissolved minerals may have leached out of the root zone.

Soil Moisture

Soil temperature impacts the physical form of water in the soil (ice, liquid, or vapor). The behavior of liquid water is also affected by soil temperature. Temperature has an impact on the density and viscosity of the water, both of which are important in determining the rate of water movement in soil. Warmer temperatures allow faster water movement in a soil.

Extreme soil temperatures could result in frozen water or excessive evaporation losses. A soil with a frozen layer of ice could impede downward movement of water, resulting in ponding of water above the ice layer. Water from spring rains on frozen ground could result in flooding of surrounding low lands due to decreased infiltration and increased surface runoff. In this situation, the surface soil is likely to erode as well.

Water expands upon freezing. Soils containing a (significant) amount of fine particles that can retain water will expand more than coarse textured soils. This expansion could result in internal pressures causing the soil particles to move away from the ice lenses formed during freezing of the soil water. In some areas where perennial crops are planted, pressure caused by freezing could push or lift the plant out of the soil, a phenomenon called **frost heave**. As a result the plant is pushed upward while the roots remain anchored, causing the roots to break away from the plant, and if the action is severe enough could result in economic losses to a farmer. Similarly, frozen soil conditions can push rocks buried in the soil upward to the soil surface.

Frozen water pockets can put uneven pressures on structures such as roads and shallow foundations, causing the structure to shift/settle in an uneven manner. This is one of the reasons footings that support foundations of buildings are placed at a depth below the frost zone in cold climates. In the higher latitudes, soils may be frozen for longer periods of the year. In far northern regions, the subsoil is always frozen. This condition, which is called permafrost, exists in much of northern Canada, Alaska, and Siberia. In recent years the permafrost has shown rapid melting of the ice layer with serious implications to plants and potentially increased decomposition rates of organic material.

Warmer soil conditions can promote evaporation. As the water vapor escapes the soil surface (drier soil), the heat stored in the water is lost from the soil and therefore it has a cooling effect on the soil. However, inputs of energy could quickly make up the energy lost via vapor with an increase in the temperature of the drier soil. Loss of water due to transpiration can also have the same impact of drying the soil and heating the soil much faster.

Fires

Sometimes extremely high soil temperatures are created by fires. The duration and intensity of the fire will determine soil temperatures and the depth to which the extreme heat is conducted. Extremely high temperature has the ability to burn the organic matter in the soil, resulting in an instant release of plant nutrients. While nutrient release may be a temporary benefit, the loss of organic matter and the destruction of microorganism in the soil may have long-term consequences. High temperature from occasional burning can also result in destruction of weed seeds and pathogenic organisms. However, some seeds that have hard coatings only germinate after being exposed to high temperatures. Sometimes soil temperatures are intentionally raised by the addition of certain chemicals that produce a heat-generating reaction in order to clean up lands contaminated with organic pollutants. The high temperature may cause the organic pollutant to vaporize.

Soil Formation/Classification

Temperature and water play an important role as active soil-forming factors. Solubility of minerals, decomposition of organic matter, moisture conditions, and competitive nature of dominant plants are related to soil temperature. The translocation of soluble materials and their subsequent accumulation in the profile leads to soil horizon differentiation. Soils on south-facing slopes in the Northern Hemisphere will tend to be shallower and contain less organic matter due to the more extreme (warm and dry) climate.

Mean annual soil temperature is generally used to group soils into various thermal regimes. The thermal regime is used for soil classification purposes. The thermal regime of a soil characterizes the type of vegetation that can adapt to the temperature conditions. For example, soils in/on wetlands compared to dry lands and of north-facing slopes versus south-facing slopes could have distinctly different thermal regimes, resulting in distinctly different vegetation and soil properties.

Factors Affecting Energy Inputs

Inputs of energy have a warming effect on soils. The most significant source of thermal energy gain is from solar radiation. The surface of the sun with a temperature of about 10,300°F (5,704°C) radiates energy to the earth. Incidental and sporadic inputs of energy may come from fires, warm rains, warm air masses blowing across the land surface, condensation of dew, biological activity, and similar events.

Atmospheric Conditions

The amount of solar energy reaching the earth's surface will depend on the relative position on earth, season of the year, atmospheric conditions, soil cover (vegetation, snow, or mulch), and other aspects. As energy from the sun radiates toward the earth, it must pass through the earth's atmosphere. On its journey through the atmosphere, solar radiation may not reach the earth's surface because it may be intercepted, absorbed, or reflected by the clouds, various gases, or particulate matter in the atmosphere. Not all of the solar radiation from the sun that reaches the top of the earth's atmosphere reaches the soil surface.

Land Cover

A portion of the solar energy that passes through the earth's atmosphere could be intercepted, absorbed, or reflected by land cover such as vegetation, snow, or mulch. *The portion of the solar energy that filters through the land cover and makes it through to the soil surface warms the soil.* But only a small portion of the solar radiation actually reaches the land surface and helps to warm the soil.

The type and density of vegetative cover influences how much solar energy reaches the soil surface. A thick crop/forest canopy/turfgrass or layer of snow/mulch shades the soil surface from incoming solar radiation and keeps the soil temperatures cool. In the spring, soils with large amounts of crop residue on the surface warm more slowly than bare soils. Bare soils warm and cool faster than soils that are covered with vegetation or snow. Crop residues or other types of mulch also reduce evaporation. Thus, not only do residues shade the soil, but by limiting evaporation, they also keep the soil moist. A moist soil warms more slowly because it takes more energy to warm water as compared to air in the soil pore spaces. These effects should be considered when modifying management practices after the adoption of minimum tillage practices.

Color of Surface Soil

Solar radiation that reaches the soil surface is either absorbed by the soil or reflected back to the atmosphere depending on soil surface conditions. The term albedo is used to describe the amount of incoming solar radiation that is reflected by a surface. Most soils reflect from 10 to 30% (albedo = 0.1–0.3) of the incoming solar radiation that reaches the soil surface. A light-colored surface will reflect much of this radiation (higher albedo), allowing lesser amounts of energy to be absorbed by the soil. Therefore, dark-colored soil surfaces gain a lot more energy than light-colored surfaces. The same soil will have a higher albedo when it is dry than when it is wet. By comparison, most vegetation reflects more solar radiation back to the atmosphere than soils (albedo = 0.2–0.3).

Slope Aspect

When the sun is directly overhead, its rays strike the soil surface at right angles and more heat is absorbed than when the sun is at a lower angle. The sun is more nearly overhead in the summer, resulting in a high level of energy (heat) absorption. In the fall, winter, and spring months, the sun appears lower in the sky and its rays strike the soil surface at a lower angle, resulting in less heat absorption. These differences are actually due to the absolute surface area over which a given amount of solar radiation is distributed. Obviously, soils that slope toward the sun can intercept more energy and thus be warmer than soils sloping away from the sun. A soil sloping to the south (in the Northern Hemisphere) warms more rapidly in the spring than one sloping to the north. Some specialty crops may be planted on the south side of an east–west row because of the exposure to more direct sunlight. The south side heats relatively rapidly, which speeds germination of seeds by as much as 3–5 days.

Incidental Sources of Energy

Heated air masses may be blown in from one area to another. When the warm air mass is blowing over a cooler soil, there will be energy gains to the soil. Rain or irrigation water that is warmer than the soil can also warm the soil. Biological activity in the upper layers of the soil adds small amounts of heat to the soil. Changes in temperature due to the incidental sources are often not as significant compared to direct solar radiation or fire.

Energy Inputs and Temperature Change

The temperature increase following an input of energy is dependent on the net amount of energy input, the heat capacity of the soil, the amount of heat transferred to the subsoil, and the amount of heat lost to the atmosphere. Energy inputs were discussed earlier in this chapter.

Heat Capacity

Heat capacity is the amount of energy it takes to change the temperature of a material by 1°C. Thus, a material that has a high heat capacity exhibits a smaller change in temperature for the same amount of energy input or energy lost. Each component of the soil (air, water, solids) reacts differently (resultant temperature change) to energy inputs. Among soil components, water has a higher heat capacity than soil solids, and soil solids have a higher heat capacity than soil air. It takes about 3,000 times more energy to warm an equal volume of water as compared to air.

A soil when wet will have a much higher heat capacity than the same soil when dry. Thus, it takes more energy to heat a wet soil than a dry soil. Likewise, wet soils compared to dry soils will show a slower decrease in temperature due to heat loss going from summer to fall to winter seasons. For this reason, wet soils are considered cold soils and they take longer to warm in the spring than dry soils. Thus, planting seeds in a wet soil must be delayed during the spring until the soil warms. Germination is delayed in cold/wet soils as well.

Heat capacities of soils vary depending on the relative proportion of solids, water, and air present. The air and water content of a soil are constantly changing, resulting in a dynamic heat capacity of the soil. Since water has an overwhelming effect on the heat capacity, evaporation, transpiration, irrigation, precipitation events, and drainage can bring about significant changes in the heat capacity of a soil. Therefore, soil moisture is a major factor in controlling the rate of temperature change. With a constant change in the soil moisture content, it makes studying the amount and rate of heat movement complex and difficult to predict.

The bulk density of a soil plays a minor role in altering the heat capacity of a soil. Compacted soils (high density) will have fewer large pores than noncompacted soils (lower density). Dry compacted soils will have a higher heat capacity than noncompacted soils (note: Fig. 7.1 is for soil thermal conductivity). However, for a wet soil, density and porosity play a minor role in altering the heat capacity of the soil.

Heat Transfer in Soils

Thermal energy is transferred as a result of a temperature difference within or between objects. Heat always flows from a warm object to a cooler one. Only the soil surface is subject to inputs of solar energy. Once the energy is absorbed by the soil at the

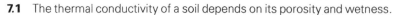

7.1   The thermal conductivity of a soil depends on its porosity and wetness.

Dry, fine-textured soils have low thermal conductivities.

Wet, coarse-textured soils have high thermal conductivities.

← Water

surface, the surface soil is attempting to reach equilibrium with the atmosphere above as well as the subsoil below. As a result, energy in the form of heat is constantly moving at all times in the soil. Subsoil temperature changes are a function of heat transfer in two directions: from the surface soil to the subsoil, and from the subsoil to the atmosphere via the surface soil. Heat can be transferred by radiation, conduction, or convection. All three types of heat transfer take place in the soil–plant–atmosphere system.

Radiation

All objects around us radiate energy in the form of invisible electromagnetic waves (short, intermediate, and long wave lengths). Solar radiation was discussed previously as the primary source of energy input into the soil. The soil surface if warmer than the air above will radiate energy to the air.

Conduction

Heat conduction occurs when energy is transferred from one molecule to an adjacent, cooler molecule. The ability of a material to conduct heat is called its **thermal conductivity**. Metals such as copper and iron have high thermal conductivities whereas materials such as wood and plastic have low thermal conductivities. The latter are called insulators. The thermal conductivity of a soil depends on the proportion of the soil volume occupied by the solid, liquid, and gaseous phases. While thermal conductance is a measure of the rate of energy movement, the resultant change in temperature is more a function of the heat capacity of the soil.

Most soil minerals have a thermal conductivity about 5 times greater than water, 10 times greater than organic matter, and over 100 times greater than air. A solid rock is able to conduct heat faster than a soil because it has neither air nor water within it. Similarly, a compacted soil or a soil with few pores can conduct heat faster than a noncompacted soil because the compacted soil has more contact between soil particles (Fig. 7.1). When a soil is wet, it has a much higher thermal conductivity than when it is dry because the air in the pore spaces is a poor conductor of heat and thus acts as an insulator. For example, when comparing a wet soil to a dry soil, it takes a lot more heat to raise the temperature of water in a wet soil, although a wet soil is able to rapidly conduct heat.

Convection

Heat is transferred by convection when the movement of a heated fluid such as air or water is involved. Furnaces that heat buildings by blowing warm air through a system of

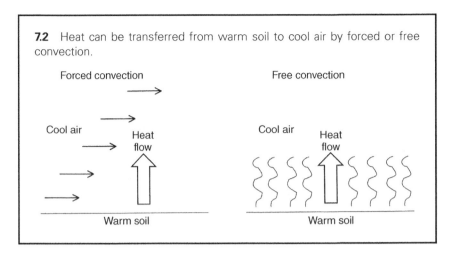

7.2 Heat can be transferred from warm soil to cool air by forced or free convection.

ducts are sometimes called convection furnaces. Because heat always flows from warmer to cooler objects, heat is transferred from a warm soil to air molecules in a cool wind. Warm air is lighter than cold air, and heat can also be transferred by convection when warm air rises. On spring mornings, sunshine on a dark, bare soil heats the soil surface and the air above it, causing the warm air to rise into the atmosphere by convection (Fig. 7.2).

Warm spring rains on cold soils could carry enough heat to instantly increase the temperature of a cold soil, but the newly gained soil water has a tendency to slow the increase of soil temperature from future energy inputs. Summer rains can also cool a warm soil. As water is drained out of the shallower parts of the soil profile, the heat stored in the water moves with the water to deeper parts of the subsoil. This kind of subsoil drainage where excess water from the root zone is removed (lower heat capacity) helps the soil warm up faster with renewed solar energy inputs.

Sensible and Latent Heat

Sensible heat is heat energy that is transported by a body that has a temperature higher than its surroundings via conduction, convection, or both. Heat stored in the soil may be transferred to the air just above the soil surface and this heat loss from the soil is considered **sensible heat**. This situation is typically noticed when going from summer into the fall season as well as after sunset.

Latent heat describes the amount of energy in the form of heat that is required for a material to undergo a change of phase (also known as "change of state"). In the soil, energy used to change ice to water is the **latent heat of fusion**, and the energy used to convert liquid water to vapor is called the **latent heat of vaporization** (Fig. 7.3). The energy in the form of heat is utilized when going from solid to liquid to gas, but heat is released when going in the opposite direction.

Latent heat is simply a shift of energy from one phase of water to another. For example, converting liquid water to vapor involves the transfer of energy from the liquid to vapor phase, resulting in more stored energy in the vapor phase. The stored energy is lost from

7.3 The surface energy budget summarizes heat flow in the soil–plant–atmosphere system. Incoming solar radiation evaporates water, warms the air, and warms the soil that emits long-wave radiation.

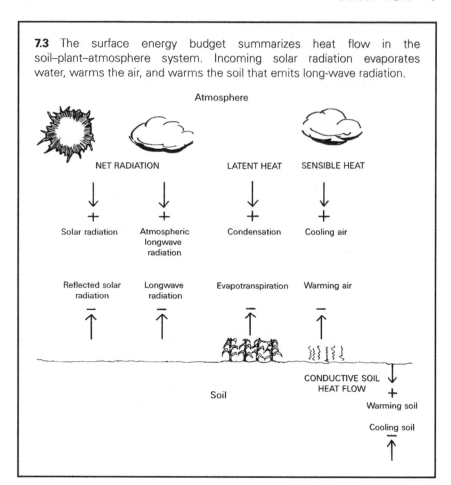

the soil only when the water vapor is physically removed from the soil by evaporation. The losses of water vapor (drier soil) results in a loss of energy, as a result one would expect cooling of the soil. Instead, a drier soil because of its lower heat capacity can rapidly increase in temperature with any gain in solar radiation.

Condensation of water vapor in the form of dew on the soil surface will release energy stored in the vapor back into the soil as the vapor converts to water. During cool nights, when the leaf surfaces have cooled enough to have water vapor in the air condense on leaf surfaces, the condensed water releases heat to the plant's microenvironment. Prolonged presence of dew on leaf surfaces causes plant pathogens to proliferate.

In temperate regions, the change of seasons from fall to winter could result in freezing of the soil moisture, the freezing process releases heat stored in the water. This heat is lost to the atmosphere, resulting in further cooling of the soil. This concept is utilized by managers of citrus orchards where they spray their trees with water before an anticipated freeze; the subsequent release of heat keeps the leaves and fruit from freezing.

There is an interaction between the latent and sensible heat transfer. When a soil is moist, the latent heat dominates energy transfer from incoming solar radiation (keeping the air above the soil cool). But when the soil dries, the amount of energy used as latent heat decreases and more energy is used to warm the soil solids with a resultant increase in sensible heat, thereby warming the air. If no rain falls for a long period, the crops growing on these dry soils may suffer not only from a lack of moisture but also from excessive heat due to increased sensible heat. Understanding sensible heat flow is important because many crops have an air temperature at which they grow best.

Soil Temperature Fluctuations

The temperature of a soil at a given time is a function of the combined effect of thermal properties of soils, atmospheric conditions, heat transfer, and net stored energy (gains and losses). The dynamic nature of energy fluctuations are reflected in daily and seasonal variations of soil temperature.

During the day due to incoming solar radiation, the surface soil warms and begins to cool off as the sun sets. The temperature of the soil at the surface is generally slightly warmer than the air above it. At midsummer a soil without vegetative cover may exhibit temperature fluctuations at the surface as much as 40°F (22.2°C) in the course of a day. As the energy must be conducted to the subsoil, the fluctuation in daily temperature is progressively less in deeper sections of the profile. For example, at a depth of 6 in. in the same soil, the variation in temperature in a day may be only 10°F (5.6°C), and at a depth of 24 in. (60 cm) the change in one day would be almost negligible (Fig. 7.4).

Depth is also a factor in the variation of soil temperature over a year. On an annual basis, the highest temperatures in the upper 1 ft (30 cm) of soil are normally reached in late summer, whereas the lowest temperatures come in late winter. At lower depths of 2–4 ft (0.6–1.2 m), high and low temperatures lag behind the surface temperatures by 2–3 months.

Depth-related soil temperatures can range widely during the year depending on which part of the world they are located. Many soils in the temperate region show a range of as much as 60°F (33.3°C). In the spring, more heat goes into the soil during the daytime than goes out at nighttime, and the soil slowly warms from day to day. In the fall, the opposite happens and less heat is conducted into the soil as the days grow shorter and cooler whereas a greater proportion of heat radiates to the atmosphere during the long nights.

Managing Soil Temperature

Since temperature has such a profound effect on the rate of biological processes in the upper few inches of a soil, considerable effort has been put into developing ways to modify the surface energy balance in horticultural and agricultural production. In managing soils for temperature, attempts should be made to create nearly optimum conditions for enhancing plant growth and productivity. Since energy is lost or gained at the soil surface, a majority of the temperature control methods focus on modifying conditions at the soil surface in hopes of restricting or enhancing movement of heat in the desired direction. Another way to manage soil temperature is through the alteration of soil moisture conditions.

Surface Conditions

Tilling the soil and creating a rough surface is likely to lower the albedo and improve heat gain, resulting in increased daytime high temperatures. Because of increased macropores in freshly tilled soils, transmission of heat to the subsoil is slower, and as a result loose, tilled soils will lose much of the heat gained during the day to the atmosphere at night.

7.4 Variations of surface and subsoil temperatures throughout the day—warming during the day, cooling at night.

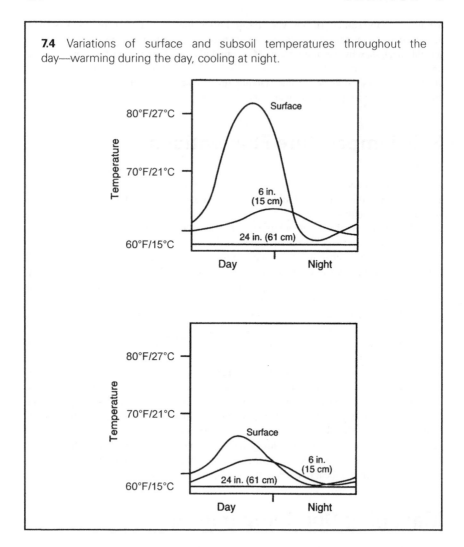

Manipulation of the surface albedo has been used to increase and decrease soil temperature. As discussed earlier, planting of vegetation has an ability to shade the soil with foliage increasing the surface albedo and therefore providing a cooling effect on the surface soil. In hot and arid regions, increasing the surface albedo by whitening the surface with a white powder can have a cooling effect on the soil. Conversely, in subarctic regions, soil temperatures have been increased by blackening crop residue to lower the surface albedo. These types of practices are only temporary and relatively expensive so they are only feasible in extreme cases or with high-value crops.

Surface slope aspect can also be modified on a small scale to increase soil temperature (Fig. 7.5). This is especially important at planting time at higher latitudes when cold, wet soils delay emergence and early seedling growth. Tillage operations can be used to create a ridge and furrow geometry in east–west rows. Soil in the ridge will warm faster because it

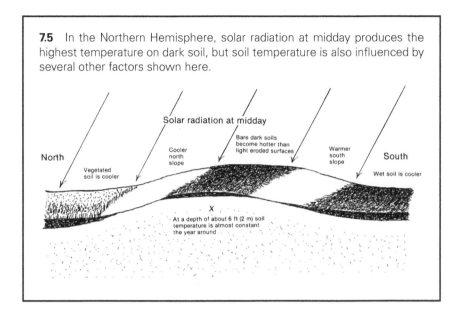

7.5 In the Northern Hemisphere, solar radiation at midday produces the highest temperature on dark soil, but soil temperature is also influenced by several other factors shown here.

dries more quickly and it absorbs more sunlight on its south-facing slope (in the Northern Hemisphere). Seeds planted into the ridge will germinate and grow faster than if they are planted in flat soil.

Mulches

Mulches applied to the soil surface with the intent of enhancing or restricting heat flow include plastic or paper sheeting; organic by-products (crop residues, leaf litter, wood chips, etc.), and gravel. All these mulches create a water and/or heat barrier at the soil surface. Porous mulches like crop residues and wood chips reduce evaporation by providing a barrier to water vapor. Mulches made of crop residue also have a higher albedo than the underlying soil. Soils under a porous mulch cover will be cooler and moister than bare soils during the growing season. Organic mulches with large particles tend to have a large volume of air between the particles. Since air is a poor conductor of heat, it serves as an insulator and keeps the heat from flowing into or out of the soil.

Clear plastic film mulches not only reduce evaporation, but also increase soil temperature by trapping long-wave radiation under the plastic and/or by decreasing the surface albedo. A dark plastic mulch will absorb more heat from the sun. Plastic mulches are common in horticultural applications where young plants require a warm, moist environment.

Soil Moisture Control

Water regulation may be the most significant temperature control method for soils. Presence of excess water can result in lower than optimum soil temperature. Excess water is common in low-lying areas or poorly drained soils. Since a drier soil will heat faster, attempts to avoid or decrease excess water content of the soil could be helpful in regulating soil temperature. To control excess water in a soil, we can avoid the entry of excess water into the soil (surface drainage) or remove the excess water that has already infiltrated into the soil profile by draining it (subsurface drainage).

Surface drainage involves the modification of land surface slopes so as to create a balance between overland flow (excess water) and the amount of water infiltrating the soil surface. Subsoil drainage requires the placement of drainage tiles/perforated tubing in the subsoil or ditches that will allow the excess water in the soil pores to freely flow into the tile/tubing or ditch. The tubing or ditch is sloped so as to carry away the excess water into a larger receiving body such as a stream. The ideal condition for soil warming is to have enough water to provide rapid heat movement to the subsoil, but not so much water as to slow the increase in soil temperature (due to the high heat capacity of water).

In nature, soil temperatures are influenced by the constantly changing meteorological conditions at the soil–atmosphere interface. External influences such as day or night, summer or winter, cloudy or sunny days, rain events, warm or cold wave events are constantly affecting the temperature of a soil. When factors such as geographic location, vegetative cover, and most importantly soil moisture conditions are added, predicting soil temperature becomes even more complicated and difficult.

CHAPTER **8**

Soil Fertility and Plant Nutrition

Water, carbon dioxide, and certain chemical elements called plant nutrients are essential for plant growth. Water is supplied by either rainfall or irrigation, carbon dioxide from the atmosphere, and the essential plant nutrients from the soil, fertilizers, or other soil amendments.

Soil Fertility

Soil fertility includes the ability of a soil to hold plant nutrients, the level of plant nutrients present, and the availability of the nutrients for uptake by plants. A soil that has a high level of essential nutrients available for use by plants is usually a productive soil if it also has sufficient soil water and if the crops are well managed. Plant nutrients exist in the soil in several different forms. They include the following:

Minerals. Examples include the feldspar group, which is the most abundant group of minerals in the rocks of the earth. Some are high in potassium and others in calcium. Nutrients are released from the minerals by weathering.

Cations or anions. These are plant nutrients that exist on the surface of clay or humus. These surfaces are called the exchange complex and are positive or negative. They attract, hold, and exchange the cations or anions (see Chapter 5).

Chemical compounds. There are many chemical compounds that form in the soil. An example would be the formation of phosphorus complexes on the surface of calcium carbonate.

Soluble ions. Numerous ions exist in the soil solution. Plants absorb a large portion of their essential nutrients from this source. This pool of nutrients is small, but can be readily replenished through cation exchange reactions and other buffering mechanisms.

Organic matter. Plant residues and organic matter contain the elements that plants require for growth. As decomposition (mineralization) occurs, these nutrients are released and can be used by plants.

Soil Science Simplified, Sixth Edition. Neal S. Eash, Thomas J. Sauer, Deb O'Dell, and Evah Odoi.
© 2016 John Wiley & Sons, Inc. Published 2016 by John Wiley & Sons, Inc.

Table 8.1   Elements required for plant growth and principal forms in which they are taken up by plants (Eash, Neal S., Cary J. Green, Aga Razvi, and William F. Bennett, eds. *Soil Science Simplified.* 5th ed. Ames, Iowa: Wiley-Blackwell, 2008. Copyright © 2008, John Wiley & Sons, Inc.)

Nutrient	Chemical symbol	Form(s) absorbed
Macronutrients		
Carbon	C	CO_2
Hydrogen	H	H_2O
Oxygen	O	H_2O, O_2
Nitrogen	N	Ammonium (NH_4^+), nitrate (NO_3^-)
Phosphorus	P	Orthophosphate ions (HPO_4^{2-}), ($H_2PO_4^-$)
Potassium	K	Potassium ion (K^+)
Calcium	Ca	Calcium ion (Ca^{2+})
Magnesium	Mg	Magnesium ion (Mg^{2+})
Sulfur	S	Hydrogen sulfate (HSO_4^-), sulfate (SO_4^{2-})
Micronutrients		
Boron	B	H_3BO_3 (boric acid), $H_2BO_3^-$ HBO_3^{-2}, BO_3^{-3}, $B_4O_7^{-2}$
Copper	Cu	Copper (cupric) ion (Cu^{2+}), copper hydroxide ion ($Cu(OH)^+$)
Chlorine	Cl	Chloride ion (Cl^-)
Iron	Fe	Ferrous iron (Fe^{2+}), $Fe(OH)_2^+$, $Fe(OH)^{2+}$, ferric iron (Fe^{3+})
Manganese	Mn	Manganese ion (Mn^{2+})
Molybdenum	Mo	MoO_4^{2-}, $HMoO_4^-$
Zinc	Zn	Zinc ion (Zn^{2+})
Co	Cobalt	Cobalt ion (Co^{2+})
Nickel	Ni	Nickel ion (Ni^{2+})

8.1   Phosphorus exists originally as a complex mineral with very low solubility. Weathering breaks it down into less complex forms, some of which can be used by plants.

	Name	Chemical form	Complexity	Availability of phosphorus to plants
↓ Weathering	Hydroxyapatite	$Ca_{10}(PO_4)_6(OH)_2$	Highly	Very low
	Dicalcium phosphate	$Ca_2(HPO_4)_2$	Moderate	Moderate
	Monophosphate ion	$H_2PO_4^-$	Simple	Readily

The availability of nutrients to be absorbed by plants varies according to the form in which they exist (see Table 8.1). As an illustration (Fig. 8.1), phosphorous is readily available to plants as inorganic orthophosphate ions ($HPO4^{2-}$ and $H_2PO_4^{2-}$). The bulk of the soil phosphorous exists as low solubility organic and inorganic phosphorous compounds that are not readily available for plant uptake. Nutrients in soil solution are quite readily available for use by plants. Those on the exchange complex are generally available for absorption by plants but not quite as readily available as those in the soil solution.

Nutrients present as complex chemical compounds, or as precipitated salts as well as those in organic matter are normally only moderately available depending to a great extent on soil water content, soil temperature, and soil pH. Those present as minerals are slowly available. They are released only as the mineral breaks down during the process of weathering (see Chapter 2 for details on weathering).

Most nutrients can be classified as readily available, moderately available, or slowly available. It is desirable to have an adequate supply of nutrients in a readily available form.

Conditions Affecting Level and Availability of Plant Nutrients

Certain soil characteristics influence the availability of nutrients. One is soil texture, the relative amount of sand, silt, and clay. Because clay particles provide a part of the exchange complex, the percentage of clay in the soil influences the capability of a soil to hold nutrients. The percentage of clay determines the size of the "nutrient warehouse."

The type of clay is also important. As explained in Chapter 5, three of the types of clay in soils in the United States are kaolinite, illite, and smectite (montmorillonite). Each type has a different capacity to hold nutrients (cation exchange capacity, CEC). In general, soils with a high CEC will be more fertile because more nutrient cations can be held on the soil complex.

The CEC is relatively low for kaolinite (3–15 $cmol_c$/kg), moderate for illite (10–40 $cmol_c$/kg), and high for smectite (80–100 $cmol_c$/kg). It follows that a soil with 20% clay as smectite would have a much greater capacity to hold nutrients than a soil with 20% clay as kaolinite. The clay content and the type of clay are both important in soil fertility.

Soil texture can also influence water retention and drainage. Sandier soils tend to drain more quickly and retain less water than do soils with higher clay contents. Sandy soils also have large pore spaces, allowing for more leaching of nutrients. The effects of soil water on nutrient availability are discussed in the following.

Structure is defined as the arrangement of soil particles into aggregates. A good soil structure is essential for water and nutrient movement and retention, as well as root growth. Large spaces between aggregates allow soil water (and the nutrients dissolved therein) to flow more freely, resulting in leaching losses. Small or no spaces between aggregates, especially due to compaction, prevent water from moving through the soil profile, resulting in runoff.

Organic matter is another important soil characteristic that, if high enough in content, can favorably impact the availability of nutrients. It has a threefold effect on fertility. First, the fraction of organic matter that is humus (the colloidal fraction) is similar to clay particles in that it has an exchange capacity ranging from 50 to 200 $cmol_c$/kg (depending on the pH of the soil) and attracts and holds nutrients for plant uptake. Second, as organic matter decomposes (mineralizes), the essential plant nutrients it contains are released and organic acids are formed that increase the availability of most nutrients. Third, an adequate level of organic matter in a soil is generally desirable not only from a plant nutrient standpoint but also because of its favorable effect on soil characteristics such as physical condition, water-holding capacity, and infiltration rate as explained in Chapter 6.

Soil water content also influences nutrient availability. Most nutrients utilized by plants are absorbed from the soil solution. A higher level of soil water normally means a higher level of most nutrients in solution, leading to improved nutrient uptake by diffusion and root interception. Adequate moisture also increases the rate of organic matter decomposition (see Chapter 5), which releases N, P, and S. Soluble and mobile nutrients such as N in the nitrate form may be lost in a process known as leaching as water percolates below the root zone. Poorly drained or very wet soils increase the solubility of minerals such as iron and manganese. Furthermore, nitrate may be lost due to denitrification (see Chapter 4) in flooded soils. Low moisture can result in reduced nutrient uptake due to formation of insoluble compounds. Higher soil temperature usually leads to greater availability of most plant nutrients, as explained in Chapter 7.

Soil pH, which is a measure of the degree of soil acidity or alkalinity, also influences the availability of nutrients through its effects on root growth and nutrient form (see Chapter 5). Most nutrients are at their highest level of availability when the pH is slightly acid to neutral (pH 6.0–7.0). Acidic soils reduce root growth, which is critical to P uptake. The pH is also important in N transformations, such as mineralization, nitrification, and N fixation, as the bacteria involved are sensitive to pH. The different forms of N have different availabilities as they have different leaching capabilities. As a soil becomes more acid, certain nutrients become less available; if a soil becomes alkaline, the availability of certain nutrients decreases (see Fig. 8.2).

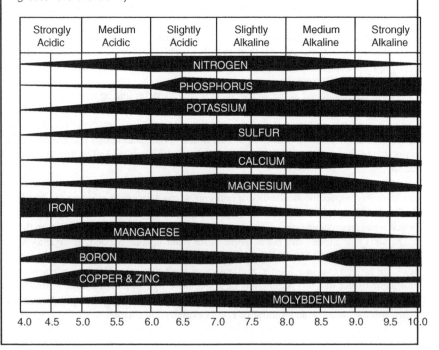

8.2 The influence of soil pH on nutrient availability. The wider the bar, the greater the availability.

Nutrient Mobility in Soils

Nutrient mobility in the soil affects the ease of its uptake by plants, and the likelihood of its leaching through the soil. Mobility of nutrients within the soil is influenced by soil physical properties such as soil texture and structure, soil chemical properties such as CEC and AEC and pH, as well as soil conditions such as moisture. Calcium, potassium, and magnesium are positively charged ions (cations). Most soil colloids have a net negative charge (see Chapter 5). Since opposite charges attract, these cations are attracted and held onto the cation exchange sites and are released only when excess cations are added to replace their place on the exchange sites. For this reason, these cations are considered less mobile and are slowly available to plants. Their movement and enrichment in waters is rarely an environmental issue, particularly where erosion is controlled.

Sulfur occurs as the anion sulfate form (SO_4^-), which does not bind to cation exchange sites and is thus mobile in most soils. Nitrogen usually exists in the soil as ammonium (NH_4^+) and nitrate (NO_3^-) forms, hence its mobility depends on the form it is in. The NH_4^+-N can be held on cation exchange sites and is therefore relatively immobile, and is not susceptible to leaching. On the contrary, the negatively charged NO_3^- ion is not held on the exchange sites, hence it is very mobile in soil water. In addition, nitrates are highly soluble in water and are subject to losses to ground water by leaching, and to surface waters by runoff (see Chapter 4).

Phosphorus normally exists in soils with pH values between 5 and 7 as the ortho-phosphate ion ($H_2PO_4^-$, HPO_4^{2-}). Unlike the nitrate anion, the ortho-phosphate ion forms very tight bonds to soil particles. As a consequence, phosphorus is typically immobile in soil, and it does not readily leach out of the root zone. The potential for P-loss is mainly associated with erosion and runoff. The lack of P mobility also reduces the availability of P-fertilizer to plants. Both nitrates and phosphorus when transported to surface waters stimulate algal growth to the point of crowding out other more desirable species through a process called **eutrophication** (see Chapter 4).

For nutrients to be utilized by plants, they must move to the surface of the plant roots, where absorption takes place. There are three processes by which nutrients move to the root surface. These are root interception, mass flow, and diffusion.

Root interception: Root interception, also known as contact exchange, occurs when the root comes into direct physical contact with nutrients associated with soil colloids as it grows through the soil. Root interception generally increases as the surface area and mass of the root increases, thereby enabling the plant to explore a greater volume of soil. Mycorrhizae also increase the surface area explored by roots thereby enhancing root interception. Root interception is an important mode of transport for calcium and magnesium. However, since the volume of soil occupied by roots is usually less than 1%, root interception is a minor pathway for nutrient transfer.

Mass flow: Mass flow occurs when dissolved nutrients in the soil solution are transported to the surface of roots as plants take up water for transpiration. Mass flow decreases as soil water content decreases. Nitrate, sulfate, calcium, and magnesium are largely supplied by mass flow.

Diffusion: Diffusion is the movement of nutrients to the root surface in response to a concentration gradient. Continued uptake of a nutrients by plants causes its concentration in the soil solution adjacent to the root surface to decrease. This creates a concentration gradient from the bulk soil to the root surface that causes nutrients to move to the root surface, where they can be taken up. Diffusion is largely responsible for supply of phosphorus and potassium.

Methods to Increase the Availability of Added Nutrients

There are a number of ways to increase the availability of nutrients. For example, potassium fertilization prior to application of ammonium fertilizers can be used to reduce NH_4^+ fixation in soils with vermiculite and illite types of clay. Early season uptake of phosphate ions by crop roots can be facilitated by placing phosphorus-containing fertilizer in or close to the seed-row at planting. In this way, phosphate ions are taken up by the roots before they react with cations dominating under acidic (e.g., Al^{3+} or Fe^{3+}) or alkaline (e.g., Ca^{2+} or Mg^{2+}) soil conditions. Under alkaline soil conditions, the phosphate fertilizer can be applied in bands with a fertilizer that generates ammonium (NH_4^+) ions. This allows slight acidification of the soil adjacent to the fertilizer band. Alternatively, compound nutrient fertilizer granules that contain nitrogen (N), phosphorus (P), and/or elemental sulfur can be applied to alkaline soils. In this case, the soil adjacent to the granule will be acidified and P uptake will be enhanced. Addition of lime to acidic soils can also enhance availability of some nutrients.

In high pH soils, soil applied iron (Fe) fertilizers often do not successfully correct Fe deficiencies. This is because the Fe^{3+} ions from the Fe fertilizer react so fast with soil that the nutrient is tied up and rendered unavailable to plants. In these soils, Fe enhanced can be corrected through foliar application of soluble iron fertilizer compounds. By avoiding the soil and applying the Fe directly to the leaves, the small amount of Fe required by plants is successfully introduced into the crop.

Plant Nutrition

Essential Elements

At least 17 elements, called plant nutrients, are essential for plant growth (Table 8.1). The first group includes three elements—carbon, hydrogen, and oxygen—that are the basic building blocks of all plant compounds. These three are needed in much larger quantities than all others combined. The initial product of photosynthesis is the simple sugar $C_6H_{12}O_6$. The carbon and oxygen come from carbon dioxide, and the hydrogen comes from water. The oxygen in the water is given off by plants and goes back into the atmosphere. This process assures us of a continuing source of oxygen.

$$6CO_2 + 6H_2O + \text{sunlight} \rightarrow C_6H_{12}O_6 + 6O_2$$

Carbon   Water   Sugar   Oxygen
dioxide

The second group of essential elements, called macronutrients, consists of nitrogen, phosphorus, potassium, sulfur, calcium, and magnesium. They are classified as macronutrients because they are used in relatively large quantities by plants.

Another group of eight elements is called micronutrients because they are normally used in smaller quantities. This group includes iron, zinc, manganese, copper, boron, molybdenum, nickel, and chlorine.

Some scientists contend that some other elements may also be essential for plant growth. Included in this group are silicon and sodium. These two elements plus vanadium, cobalt, and iodine are often called beneficial elements because they can be used by plants as substitutes for nutrients that are essential.

Approximately 90% of the dry weight of a plant is made up of carbon, hydrogen, and oxygen; the balance consists of the other essential elements. Most of this remainder consists of the elements classified as macronutrients, whereas less than approximately one-tenth of this 10% is in the micronutrient group. Several elements not known to be essential may also be included in the 10%.

Natural Sources of Plant Nutrients

Plants obtain carbon, hydrogen, and oxygen from soil water and atmospheric carbon dioxide and assimilate them into glucose during photosynthesis. The remaining 14 elements are normally absorbed in ionic form from the soil (Fig. 8.3, Table 8.1), although small amounts can be taken in through the leaves if placed there in solution by precipitation, foliar application, or sprinkler irrigation. Nitrogen comes originally from the atmosphere, which is nearly 79% N, and is in a form that plants cannot use.

Nature has several methods by which atmospheric N is converted into forms that plants can use. They include the following:

Bacteria and leguminous plants join together in a process called symbiotic nitrogen fixation, which, when combined with other steps in the nitrogen cycle, provides nitrogen in a form usable by plants (see Chapter 4).

Fixation of nitrogen by **soil bacteria** without the help of legumes called nonsymbiotic fixation.

The action of *lightning discharging* in the atmosphere causes nitrogen oxides to be formed, which are then brought to earth by rain. These latter two sources of nitrogen provide relatively small quantities for plant growth.

Within soils, most N is in the form of organic matter. While this pool of N is very important, N in the organic form generally is not available to plants. As discussed in earlier chapters, the organic N must undergo mineralization to be made available for plants.

The remaining 13 essential elements are naturally derived from the weathering of rocks and minerals of the earth. Phosphorus, for example, comes principally from a mineral called apatite, whereas magnesium comes from minerals such as serpentine and dolomite. When rocks and minerals undergo weathering, elements are released and become part of the soil system.

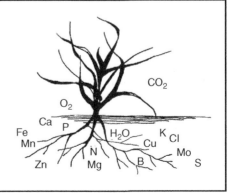

8.3 Carbon and oxygen come from carbon dioxide in the air, hydrogen from water in the soil, and other elements are absorbed by plants from the soil.

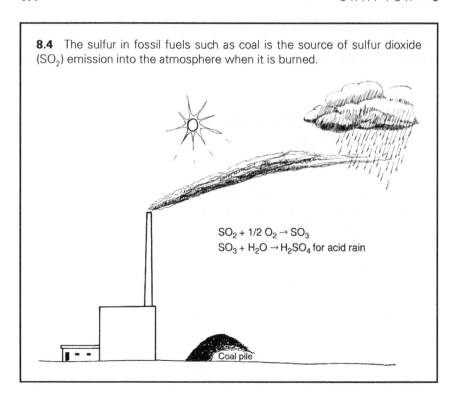

8.4 The sulfur in fossil fuels such as coal is the source of sulfur dioxide (SO_2) emission into the atmosphere when it is burned.

$$SO_2 + 1/2\ O_2 \rightarrow SO_3$$
$$SO_3 + H_2O \rightarrow H_2SO_4 \text{ for acid rain}$$

Coal pile

Sulfur is often emitted into the atmosphere by coal-burning facilities as sulfur dioxide, which is moved by air currents and then carried to the soil by precipitation (Fig. 8.4). This can also be true for a small portion of the nitrogen used by plants. Plant nutrients can also be added through irrigation water.

Role of Essential Plant Nutrients

Each plant nutrient plays one or more special roles in plant growth. Table 8.2 lists each nutrient, one of its functions in plant growth, and some deficiency symptoms. A nutrient may be the essential part of a plant compound, thus providing it with a structural base. Calcium, for example, is part of calcium pectate, which is a compound that is a part of the plant cell wall.

Other nutrients may be essential for making compounds involved in plant growth processes, such as phosphorus as a part of adenosine diphosphate (ADP) and adenosine triphosphate (ATP), which are two compounds involved in the transfer of energy within a plant. The compounds used as storage of plant foods such as protein require nitrogen and sulfur.

Another group is involved in the regulation of certain enzymatic processes. Enzymes in plants such as catalase and lactase act as catalysts or activators. They often contain micronutrients such as iron and copper.

Table 8.2 Essential plant nutrients, function in plant growth, and deficiency symptoms (Eash, Neal S., Cary J. Green, Aga Razvi, and William F. Bennett, eds. *Soil Science Simplified.* 5th ed. Ames, Iowa: Wiley-Blackwell, 2008. Copyright © 2008, John Wiley & Sons, Inc.)

Plant nutrient	Function in plant	Deficiency symptom
Nitrogen	Essential part of amino acids, protein, and chlorophyll	Yellowing of midrib of lower leaves
Phosphorus	Part of energy transfer compounds	Reddish-purple color of leaves of young plant
Potassium	Regulation of osmosis and water use and transportation system	Browning of outer edges of lower leaves
Calcium	Formation of calcium pectate used in cell walls	No development of terminal buds and apical tips of roots
Magnesium	Central atom of the chlorophyll molecule	Interveinal chlorosis of middle or lower leaves
Sulfur	Essential part of three amino acids essential for protein formation	Uniformly chlorotic upper leaves and slow growth
Iron	Component of chlorophyll and cofactor for enzymatic reactions	Interveinal chlorosis in young leaves
Zinc	Involvement in auxin metabolism and part of dehydrogenase enzyme	Spotted white or yellow areas between veins of upper third of leaves; also lack of terminal growth
Manganese	Electron transport and part of enzyme system	Interveinal chlorosis in young leaves
Copper	Part of oxidase enzyme system	Yellowing and stunting of young leaves
Nickel	Involved in enzyme converting urea to ammonium	Interveinal chlorosis in young leaves—progressing to necrosis
Molybdenum	Part of nitrate reductase enzyme	Yellowing of midrib of lower leaves
Boron	Growth and development of a new meristematic cells	Pale green young leaves; leaves die and terminal growth ceases
Chlorine	Osmotic and cation neutralization	Partial wilting and loss of leaf turgor when moisture is adequate

Determining Nutrient Needs

To be able to produce top yields, all essential plant nutrients must be present in adequate quantities. The most common element to be deficient for most crops and lawns is nitrogen. Phosphorus is normally the second most common element to be deficient. Potassium, calcium, and magnesium are often lacking in soils in the eastern half of the United States where rainfall normally exceeds 25 in. (625 mm) per year; whereas in the western United States, these elements are well supplied in most soils.

Several methods are available to determine if a nutrient is deficient and the quantity needed to correct the deficiency. These methods include chemical analyses of the soil and the plant, nutrient deficiency symptoms, and growth tests. In order for chemical analysis of the soil and plants to be of value, soil and plant samples need to be representative of the area in question.

Soil Sampling

The method most often used for determining nutrient need is chemical analysis of soil. To determine nutrient need most accurately, two things are required: (1) a soil sample that truly represents the field in question, and (2) the chemical method that has been adequately researched and properly correlated/calibrated for the crops and soils in question.

For soil sampling, there are three common approaches. They are sampling (1) by soil type, (2) on a grid basis, or (3) on a management zone basis, which determines sampling locations based on yield maps, remote sensing, past management history, and so on. The same basic principles apply to sampling lawns and gardens but on a smaller scale.

To sample by soil type, diagram a field by soil type (such maps are available in soil surveys) and obtain a composite sample from each soil type (see Fig. 8.5). For each composite sample, systematically or randomly take 10–15 individual cores of one soil type using a sampling tube or a shovel. A soil core covers one inch square area and may extend to varying soil depths, depending on the type of the crop to be grown (shallow rooted vs deep rooted), the field (pasture, no-till, tilled), and type of test (mobile nutrients vs immobile nutrients) to be performed. Usually, soil cores are taken to a depth of 6 in. (15 cm) in pasture, turf, no-till land, tilled fields, or gardens. Testing for mobile nutrients, such as nitrates and sulfates requires deeper sampling. Take samples with stainless steel or chrome-plated sampling tools and clean plastic buckets to avoid contaminating the samples with micronutrients, particularly copper and zinc. Do not use brass, bronze, or galvanized (zinc plated) tools.

Thoroughly homogenize the cores and remove approximately 1 pint (0.5 l) or 1 lb. (0.4 kg) to submit for testing. Repeat the same process for each soil type in the field. For fields larger than 12 acres (5 ha), proportionately more cores should be taken. The more the cores taken, the more reliable the measure of fertility of the field. One sample should not represent more than 25 acres (10 ha). If one soil type area in a field is too small to be fertilized separately (e.g., less than 5 acres, or around 2 ha, in size), do not sample any of the area.

If the samples are taken systematically, the area sampled should be traversed in a zigzag pattern (See Fig. 8.5, step 2) to provide a uniform distribution of the sampling sites in the entire area. High variation within the field being sampled will decrease the accuracy and reliability of a soil test recommendation. For this reason, a uniform soil should not differ in soil type, color, slope, drainage, texture, past management, and natural vegetation, and any parts of the field showing these differences should be sampled separately. Areas with recent lime, manure, composts, and fertilizer additions should not be sampled. Severely eroded sections, manure or lime stockpile areas, wet spots, old building sites, fence rows, and areas adjacent to gravel roads should also not be sampled. Soil samples taken from these locations would not be typical of the soil in the rest of the field, hence including them would produce misleading results.

To sample on a grid basis, divide a field into 3–5 acre (1.2–2.0 ha) squares as a grid. Take one composite sample (of 8 single-sample core samples) from each square (see Fig. 8.6.). The grids may be based on yield maps, and so on, as discussed previously. These sampling areas are not necessarily rectangular.

For grid sampling, various patterns of sampling can be used by shifting from the center of the grid to randomize the sites. This approach to sampling is being used for the computerized application of nutrients or variable-rate application, often called precision or site-specific nutrient management programs.

Ideally, soil samples should be taken as close as possible to planting or to the time of crop need for the nutrient (2–4 weeks before planting or fertilizing the crop). This is because nutrient concentrations in the soil fluctuate with the season. However, to give sufficient time to plan and implement land management decisions before the busy planting season, it is more practical to collect soil samples 3–6 months prior to planting time. Do not collect samples when the soil is too wet or too dry as it will be difficult to mix the cores. As a rule, if the soil is too wet or too dry to plow, it is too wet or dry to sample.

8.5 Proper collection of soil samples is extremely important. Tests made on carelessly taken samples can be misleading and costly. *Source:* (Eash, Neal S., Cary J. Green, Aga Razvi, and William F. Bennett, eds. *Soil Science Simplified.* 5th ed. Ames, Iowa: Wiley-Blackwell, 2008. Copyright © 2008, John Wiley & Sons, Inc.)

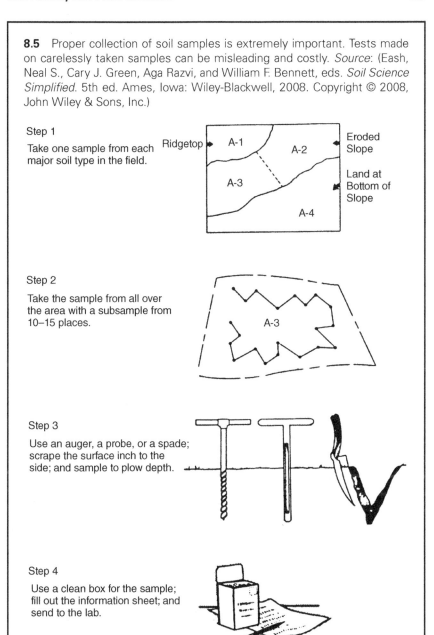

Step 1

Take one sample from each major soil type in the field.

Ridgetop — A-1 — A-2 — Eroded Slope

A-3 — Land at Bottom of Slope

A-4

Step 2

Take the sample from all over the area with a subsample from 10–15 places.

A-3

Step 3

Use an auger, a probe, or a spade; scrape the surface inch to the side; and sample to plow depth.

Step 4

Use a clean box for the sample; fill out the information sheet; and send to the lab.

To ensure accurate results and minimize changes in nutrient levels caused by soil organisms, soil samples should be handled with great care. Moist soil samples should be stored in a cool box during sampling and in a refrigerator after sampling. If it is not feasible to refrigerate or freeze the samples soon after collection, take them to the soil testing

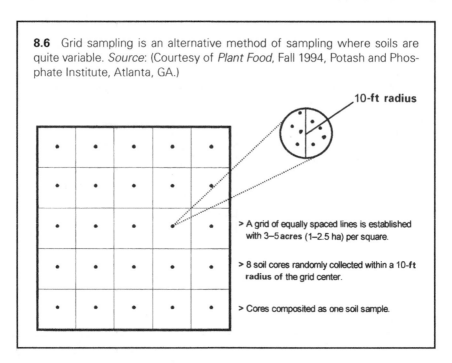

8.6 Grid sampling is an alternative method of sampling where soils are quite variable. *Source*: (Courtesy of *Plant Food*, Fall 1994, Potash and Phosphate Institute, Atlanta, GA.)

10-ft radius

> A grid of equally spaced lines is established with 3–5 acres (1–2.5 ha) per square.

> 8 soil cores randomly collected within a 10-ft radius of the grid center.

> Cores composited as one soil sample.

laboratory or air dry them by spreading the soil on a plastic sheet. Break up all clods or lumps and dry at room temperature. A circulating fan may be used to facilitate drying.

Take soil samples at the same time of the year, once every 2 or 3 years. Sampling at the same time of year minimizes the effect of seasonal variations on soil test results.

Soil Test Methods

Composite samples should always be placed in a clean container to avoid contamination. It is best to use containers provided by the soil-testing laboratory, if available. Be sure to provide the producer's name, address, and field number for each sample. Also provide information on cropping history for at least 2 years, previous fertilizer use and manure applications, and yield levels, crop to be fertilized, and anticipated yield potential for the next crop. Be sure to choose a well-qualified laboratory whose agronomist is familiar with the soils and crops in the producer's locale. State-operated soil-testing laboratories vary in their specific instructions for soil sampling so it is advisable to check with the local county agricultural extension agent.

A soil test is a chemical assessment of the nutrient-supplying ability of a soil at the time of sampling. Many methods of soil testing are available. Some measure the total content of a nutrient in the soil, while other tests attempt to measure the "available" nutrient levels. Most testing that is done to predict fertilizer needs for a crop is in the second category, which is to provide an index, or an estimate of the nutrient-supplying ability of the soil. During testing, this fraction is separated from the soil using an extracting solution that is mixed with the soil for a specified period of time. The nutrient in the extractant is then analyzed after filtration.

For any method to work, the soil tests have to be evaluated on the basis of actual crop response. Without locally applicable crop response data, a soil test is useless. Each test must be correlated/calibrated with long-term field experiments and fertilizer trials so as to assess the likelihood of a yield response to addition of fertilizer. For the calibration to be complete, these soil test correlation trials must be conducted for several years on a specific crop growing on a specific soil type.

Soils that contain a high amount of available nutrients as indicated by high soil test values require less fertilizer input than do soils that contain a low amount of available nutrients. A low soil-test value for a particular nutrient indicates that the crop will not obtain enough of that nutrient from the soil to produce the highest yield under average soil and climatic conditions. Supplementation through fertilizer addition will be necessary to correct nutrient deficiency. In general, there is a high chance of getting a response to a nutrient if the soil test is low.

The test is then calibrated to determine the amount of each nutrient needed to maximize profit from fertilizer application. The soil test level above which crop yields remain the same even though soil fertility continues to increase is known as the critical soil test. In other words, no response is expected above the critical soil test level and a response is expected below it. Nutrient guidelines for fields testing below the critical soil test value are determined by conducting yield trials on a number of fields, across a full range of soil test levels. The treatments on each field are selected to represent the full range of rates of the nutrient in question (e.g., 0, 30, 60, 90, and 120 lb or 0, 12, 24, 36, and 48 kg of P_2O_5/acre). The results from these tests indicate how many pounds of fertilizer are needed at a given soil test level to reach economically optimum yield, that is, the fertilization level that brings the most profit per acre (i.e., the value of extra yield vs the cost of extra fertilizer) at a given soil test level. Therefore, by using the best soil-testing procedures and sound fertilizer recommendations based on adequate field research, laboratories are able to predict the optimum economic rate of fertilizer that is environmentally sound.

Potential yields vary; hence, the relationship between soil test values and crop response will vary. This is because yield is affected by climate, disease, and weeds as well as soil fertility. The interpreter of the soil test results should also consider (and should be knowledgeable about) potential yield levels for any given area or even specific farms, if possible. Previous yield history, cropping systems, and fertilizer practices, if known, are needed for best recommendations.

Fertilizer Recommendations: Approaches and Philosophies

Although soil testing is the basis for determining the adequacy of many nutrients, soil tests do not provide the final answer of what fertilizer rate needs to be applied to an individual field in a given year. There are two approaches commonly used for giving fertilizer recommendations for phosphorus and potassium. These are the build-maintenance and nutrient sufficiency approaches.

Build-Maintenance Approach. The strategy here is to maintain soil fertility for future years by applying more nutrients than the crop removes, so that yields are not limited by nutrient level in the soil. To this end, enough fertilizer is applied to both meet the nutrient requirements of the immediate crop and to build up the level of nutrient in the soil to a critical soil test level over a few years. The critical soil test level is the soil test level at which near maximum (90–95%) yield is obtained. It is based on yield response curves, which are the result of years of research and trials, and are specific to a particular soil and climatic conditions.

Once the soil test value has been raised to the critical level, the soil is largely capable of supplying crop nutrient requirements in a given year and soil test level is maintained at, or above, the critical level by applying fertilizer rates to replace only the amount of nutrients expected to be removed by the crop. This keeps the soil test level from falling below optimum between soil tests. Once the soil test reaches a level where crop removal will not reduce the soil test level to below optimum, no additional nutrients are added except for the small amounts supplied in starter fertilizer applications.

Since nutrient availability in the soil is increased over time, for future years, more fertilizer is used. While this reduces the risk of nutrient deficiencies related to their availability in soil, profitability in a given year is decreased. This approach also increases the risks of over-fertilizing and negative impacts on the environment.

Since plant nutrients rarely work in isolation, over-fertilizing can also lead to antagonisms, whereby high levels of one nutrient may influence the uptake of another. For instance, excess calcium levels can cause potassium, boron, or magnesium deficiencies in some soils, excess magnesium can reduce potassium uptake and vice versa, and excess phosphorus can lead to reduced zinc uptake. Therefore, balanced fertilization is important for increasing crop yields and improving nutrient use efficiency.

Sufficiency Approach. This approach utilizes the limiting factor concept to make nutrient recommendations. The limiting factor concept states that crop yield increases will cease when a nutrient or factor "runs out;" that is, it cannot promote further increases. When that factor is supplied, yield will increase until another factor becomes limiting. Fertilizers are applied only to meet the nutrient requirements of the crop. The goal is to maximize profitability in a given year, while minimizing fertilizer applications and costs. No consideration is given to future soil test values. When soil test levels are low, crops are likely to respond to additional nutrients and the recommended fertilizer rates exceed nutrient removal by the crop. As soil test levels increase to the critical soil test level, fertilizer recommendation decrease to almost zero as these soils are unlikely to demonstrate any yield response to added fertilizer nutrients. Overall, there is decreased fertilizer usage with this approach leading to a slower increase in soil test values compared to the build-maintenance approach. Most laboratories and universities in the US use this approach for their fertilizer recommendations.

Soil Testing for Nitrogen. Nitrogen exists in organic and inorganic forms in the soil, and interconverts readily among those forms, causing large variations in inorganic N concentrations. For this reason, soil testing for nitrogen has not proven useful, particularly in the more humid climate of the Corn Belt and Southern/Southeastern United States. In these areas, weather-induced variations in inorganic N concentrations greatly affect the ability of a soil test to accurately predict N availability to the crop in a given growing season. So, while a soil test is one of the best methods to determine the right rate of P, K, and several other nutrients in these higher-rainfall areas, N fertilization programs are generally not based on pre-plant soil-testing. Rather, nitrogen management depends on knowledge of nitrogen requirements of the various crop species.

Plant Sampling

The procedure for collecting plant samples for chemical analysis is similar to collecting soil samples. First, determine whether the crop in a field is relatively uniform. Then select a plant part (or parts) from about 15 places in the field. If the grower is concerned about a problem area, take one sample (15 subsamples) of fresh material from the affected area and another sample from the area of healthy/normal growth. Samples should be taken as soon as the problem appears. These two plant samples will provide a comparison of the two areas and an indication of whether nutrient supply is adequate for optimum growth.

Plant part to be sampled depends on many factors—age of plant, type of plant, nutrient to be tested, and so on. Take care not to contaminate the samples with soil as even a small amount of soil will cause the results to be invalid. Contact your local farm advisor or laboratory consultant for information on how to sample a specific crop.

A report from the testing laboratory provides results on the soil and/or plant tests and normally gives recommendations on the type and quantity of nutrients that need to be added.

Nutrient Deficiency Symptoms

Nutrient deficiency symptoms in plants may be seen as poor growth, lack of green color (chlorosis), or browning of tissue (necrosis). The best-known nutrient deficiency is the one caused by lack of nitrogen. On a corn plant, for example, the green tissue along the midrib of the lower leaves turns yellow. The type of symptom and its location on the plant suggest the nutrient that is deficient. See Figure 8.7 for common deficiency symptoms for various nutrients. Table 8.2 also lists common deficiency symptoms for each plant nutrient.

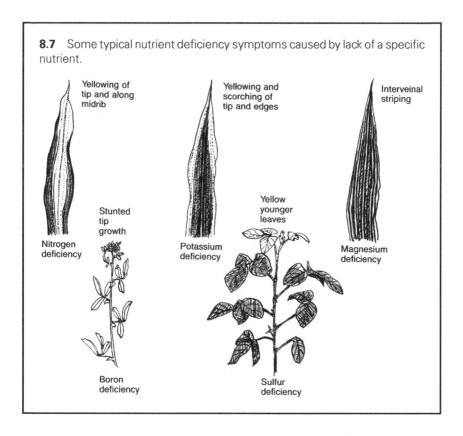

8.7 Some typical nutrient deficiency symptoms caused by lack of a specific nutrient.

Yellowing of tip and along midrib

Nitrogen deficiency

Stunted tip growth

Boron deficiency

Yellowing and scorching of tip and edges

Potassium deficiency

Yellow younger leaves

Sulfur deficiency

Interveinal striping

Magnesium deficiency

Nutrient Mobility Within the Plant

Certain nutrients, when deficient in the plant tissue, have the ability to move from older leaves to younger leaves where they are needed for growth. Nutrients with this ability are said to be mobile, and include nitrogen, phosphorus, potassium, magnesium, and molybdenum. Other nutrients do not have the ability to move from old to new growth and are said to be immobile. Immobile nutrients include calcium, sulfur, boron, copper, iron, manganese, and zinc. Knowing whether a nutrient is mobile or immobile can provide us with clues when diagnosing deficiency symptoms. If the deficiency symptoms show up on the younger, new growth, we know that the deficient nutrient is immobile. On the other hand, if deficiency symptoms appear in older mature leaves, we know that the deficient nutrient is mobile.

Deficiency symptoms should not be solely relied upon for making fertilizer recommendations. This is because deficiency symptoms may be caused by other factors including insect damage, diseases and many physiological problems. In addition, although the law of the minimum stipulates that the nutrient in the shortest supply will be the first to limit growth, other nutrients may be deficient. Collecting both soil and tissue samples from both "poor" and "good" areas of a field is the best way to diagnose nutrient deficiencies.

There are other diagnostic tools for assessing the nutrient status of crops. These include chlorophyll meters, leaf color charts and on-the-go sensors.

Biological Growth Tests

Biological growth tests may also be used to determine nutrient needs. A simple method is to split a field and apply one type of nutrient on one half and another nutrient on the other half. Or try one rate of a nutrient on one half and double the rate on the other half.

Biological tests can also be used in greenhouses for a short growth period. For such "pot" tests, up to a gallon (4 l) of soil is brought into the greenhouse and divided into small, 1-pint (0.5-l) containers, which receive various rates of the nutrients in question (leaving one untreated). Rapid-growing plants (such as small grains) are planted, harvested in a short time (such as 30 days), and weighed to determine which rate provided the greatest growth.

Adding Plant Nutrients

If the level of essential plant nutrients in the soil is low or their availability is decreased for some reason, nutrients need to be added to achieve good crop yields. Unless nutrients removed in harvested grain and plant biomass are replaced, soil fertility will deteriorate and crop yields will decline.

Fertilizers

Each essential plant nutrient (except carbon, hydrogen, and oxygen) can be applied as a commercial fertilizer. There are many different types and forms—so many, in fact, that it would be difficult to describe all of them here. But a few are discussed next.

Fertilizers come in dry, liquid, and gaseous forms. Some contain only one essential nutrient, whereas others contain two or more. Percentages of the essential nutrients in a fertilizer also vary widely. The percentage of a nutrient or nutrients in a fertilizer is guaranteed to be at a minimum level or above as required by state laws.

A nutrient guarantee is expressed in three numbers, such as 20-10-5. The first number (20 in the example) represents the percentage of available nitrogen (as N). It may be in

Table 8.3 Fertilizer grades (Eash, Neal S., Cary J. Green, Aga Razvi, and William F. Bennett, eds. *Soil Science Simplified*. 5th ed. Ames, Iowa: Wiley-Blackwell, 2008. Copyright © 2008, John Wiley & Sons, Inc.)

Grade	N	P_2O_5	K_2O
	(%)	(%)	(%)
20-10-5[a]	20	10	5
0-20-20[a]	0	20	20
10-30-10[a]	10	30	10
0-0-60[b]	0	0	60
32-0-0[a]	32	0	0

[a] Mixed fertilizers with two or more nutrients.
[b] Straight fertilizers with only one nutrient.

nitrate (NO_3^-), ammonium (NH_4^+), or organic form. The percentage of the nitrogen in each form must appear on the label. The source of any organic nitrogen also must be shown. The form of nitrogen is important in fertilizer timing as most plants use nitrates preferably and the microbial conversion of other forms to nitrates requires time. The second number (10 in the example) is the percentage phosphorus content expressed as phosphate (P_2O_5). The actual phosphorus content of the material can be calculated by multiplying the P_2O_5 by 0.44. To convert back to P_2O_5, multiply phosphorus by 2.29. The third number is the percentage potash (K_2O) content (5 in the example), which is an expression of the potassium in the material. To convert to elemental potassium (K), multiply the K_2O number by 0.83. To convert back to K_2O, multiply the potassium number by 1.20. Other examples of grades are given in Table 8.3. If there are other guaranteed nutrients present, they are listed as additional numbers with the symbols for the elements. For example, if the 20-10-5 given here also contains 2% zinc and 1% manganese, the grade would show as 20-10-5 + 2% Zn + 1% Mn. This grade guarantee is always listed on the fertilizer container (whether it is in a sack, box, or bottle) and also on the invoice. The manufacturers of the fertilizer make the guarantee and it is enforceable by state law. This normally assures the producer of purchasing the correct product.

The percentage of a nutrient in a fertilizer is important because it determines the amount to use per acre to obtain a given quantity of a needed nutrient. For example, a fertilizer with a 20-5-10 grade is 20% N, 5% P_2O_5, and 10% K_2O by weight. If 100 lb (45 kg) of this fertilizer were applied evenly over an acre, the amounts N, P_2O_5, and K_2O applied would be:

$$20\% \text{ N} \times 100 \text{ lb fertilizer/acre} = 20 \text{ lb (9 kg) N/acre}$$

$$5\% \text{ P}_2\text{O}_5 \times 100 \text{ lb fertilizer/acre} = 5 \text{ lb (2.7 kg) P}_2\text{O}_5\text{/acre}$$

$$10\% \text{ K}_2\text{O} \times 100 \text{ lb fertilizer/acre} = 10 \text{ lb (4.5 kg) K}_2\text{O/acre}$$

The concentration of nutrients in the fertilizer is multiplied by the amount of the fertilizer material applied per acre to find the actual amount of N, P_2O_5, and K_2O applied per acre. If crop P guidelines calls for 15 lb (6.75 kg) P_2O_5/acre, the amount 20-5-10 fertilizer needed would be:

$$(15 \text{ lbP}_2\text{O}_5\text{/acre})/(0.05) = 300 \text{ lb/acre of 20-5-10 fertilizer.}$$

The amount of 20-5-10 fertilizer material to be applied per acre is calculated by dividing the crop P_2O_5 requirement by the concentration of P_2O_5 in the fertilizer.

A fertilizer that contains only one nutrient is called a straight fertilizer (or a fertilizer material). An example would be ammonium nitrate, which contains only nitrogen as a nutrient. Mixed fertilizers containing two or more nutrients would be a mixture of two or more straight fertilizers. An example would be mixing ammonium nitrate (33-0-0) with a calcium phosphate (0-46-0) to produce a grade of fertilizer that might be 16-20-0. Other terms often used to describe fertilizers are complete and balanced. A complete fertilizer, a term of little significance from a crop production standpoint, is one that contains all three of the primary nutrients and would be a grade such as 24-10-8 (Fig. 8.8).

A *balanced fertilizer* contains equal amounts of N-P-K (10-10-10). A *balanced fertilizer program* is one which provides nutrients based on crop needs and nutrient deficiencies. It might include, for example, only nitrogen if it is the only nutrient deficiency. Or if nitrogen, phosphorus, potassium, zinc, and boron are needed, the fertilizer that would provide the required nutrients for a balanced fertilizer program might be a 15-5-10 + 1% Zn and 0.5% B.

Nitrogen, the fertilizer nutrient used in greatest quantities, comes in all three forms of dry, liquid, or gas. Anhydrous ammonia (NH_3) is the principal source of nitrogen used in the United States. It is in a gaseous form when applied to the soil (Fig. 8.9) but is stored as a liquid when under pressure or at low temperatures. It contains only one nutrient and its nitrogen content is 82%. This is the highest nutrient concentration in any commonly used fertilizer. Ammonia is the base for producing other nitrogen fertilizers, such as the examples given in Table 8.4.

8.8 A complete commercial fertilizer is reported in terms of varying percentages of N, P_2O_5, and K_2O.

After ammonia is manufactured, it is mixed with various acids in liquid form and the resulting product is then dried to the solid form (Fig. 8.10). While in the liquid state, urea and ammonium nitrate are often mixed to form a solution containing 28 or 32% nitrogen. This is the second most widely used source of nitrogen in the United States (Fig. 8.11).

Phosphorus fertilizers are derived from a mineral called apatite, which is a calcium phosphate and is a form in which the phosphorus is not readily usable by plants. The mineral is mined from deposits just below the surface of the soil in Florida and Idaho in the United States and in Morocco and the former Soviet Union.

Apatite is commonly called rock phosphate (Fig. 8.12). It is treated with an acid (either sulfuric or phosphoric) to produce a calcium phosphate (either 0-20-0 or 0-46-0) in which the phosphorus is in a more usable form than in apatite. Phosphoric acid also can be produced from the apatite and can then be treated with ammonia to produce an ammonium

8.9 Nitrogen may be applied as anhydrous ammonia (NH_3) gas fed from a pressure tank through hollow knives that cut into the soil.

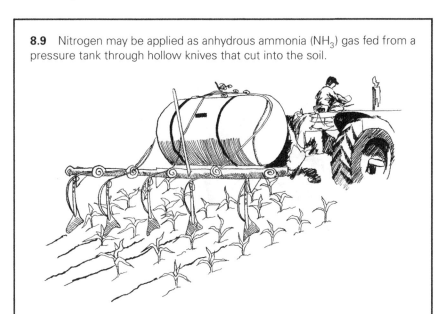

Table 8.4 Combination used to produce nitrogen fertilizers (Eash, Neal S., Cary J. Green, Aga Razvi, and William F. Bennett, eds. *Soil Science Simplified.* 5th ed. Ames, Iowa: Wiley-Blackwell, 2008. Copyright © 2008, John Wiley & Sons, Inc.)

Combinations	Product	Percentages ($N-P_2O_5-K_2O$)
$NH_3 + HNO_3$	NH_4NO_3 (ammonium nitrate)	33.5-0-0
$NH_3 + H_2SO_4$	$(NH_4)_2SO_4$ (ammonium sulfate)	21-0-0
$NH_3 + H_3PO_4$	$NH_4H_2PO_4$ (ammonium phosphate)	11-48-0
$NH_3 + CO_2$	$(NH_2)_2CO$ (urea)	45-0-0

Note: NH_3 = ammonia, HNO_3 = nitric acid, H_2SO_4 = sulfuric acid, H_3PO_4 = phosphoric acid, CO_2 = carbon dioxide.

phosphate (Table 8.4). Phosphorus fertilizers are available either in the liquid form or in the dry form. Phosphorus is second to nitrogen in quantity used in the United States.

Potassium fertilizers are manufactured from minerals such as muriate of potash or langbeinite, which occur in deposits in the earth. Some deposits such as those in Canada are fairly shallow, whereas others such as those in New Mexico are quite deep (Fig. 8.13). Muriate of potash is refined to produce potassium chloride (0-0-60); langbeinite is used to produce potassium-magnesium sulfate (0-0-22 + 11% Mg). Other common potassium fertilizers are potassium sulfate (0-0-50), potassium nitrate (13-0-44), and potassium phosphate (0-26-26).

8.10 Most nitrogen fertilizers start with ammonia, which reacts with various acids. They exist in gaseous, dry, or liquid forms.

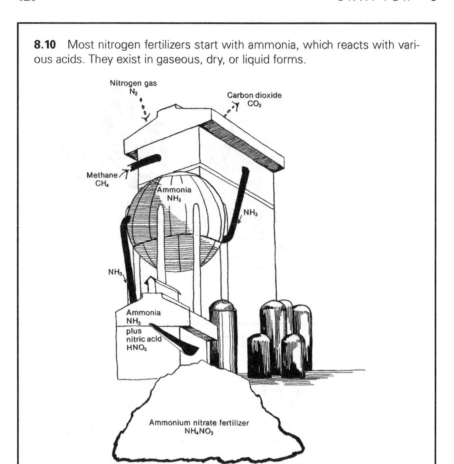

The other fertilizer elements come from various sources. Calcium, for example, comes mostly from limestone (calcite) and gypsum. Magnesium is used either as potassium-magnesium sulfate, limestone (dolomite), or magnesium sulfate. Sulfur is usually applied as elemental sulfur, a thiosulfate, or as one of the sulfate forms such as ammonium or potassium sulfate.

The micronutrients iron, zinc, manganese, and copper are usually used in one of three forms. A sulfate salt such as zinc sulfate is common. Another is called a chelate, which is an organic form that reacts with the micronutrient to make a relatively soluble product. A third form is an oxide, such as zinc oxide.

It is important to use the source of fertilizer best suited for any given crop and condition. A local fertilizer dealer, consultant, or agricultural agent should be consulted for specifics on the best one to use.

8.11 Liquid fertilizer may be applied to the soil or, if sufficiently diluted, it can be used as a foliar application.

8.12 Rock phosphates for making fertilizer are mined from open pits.

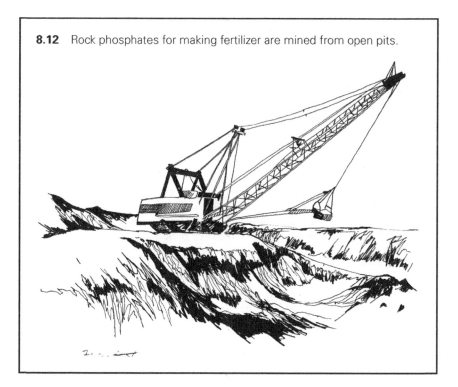

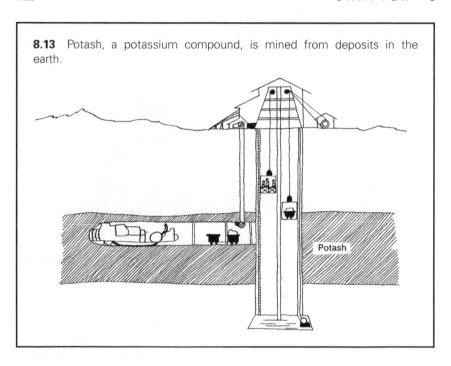

8.13 Potash, a potassium compound, is mined from deposits in the earth.

Potash

Effect of Fertilizers on Soil pH

Different fertilizer materials have different impact on soil pH. Generally, fertilizers with high proportions of total nitrogen and are derived from ammonium sources (such as urea, ammonium sulfate, ammonium phosphate, or ammonium nitrate) can acidify soils with repeated applications. Most fertilizers provide the "Lime Equivalent" on the bag's label. The lime equivalent is the amount of limestone (calcium carbonate) it takes to neutralize the acidifying effects of using one ton of a particular fertilizer.

Some fertilizers can also increase soil pH. These fertilizers are usually low in ammonium, but high in nitrate. Additionally, these fertilizers sometimes contain calcium from calcium nitrate. The lime equivalent for these fertilizers is also given, but it indicates the equivalent liming effect rather than the lime needed to offset acidity.

The following generalizations may be used as a guide:

- Ammonium (NH_4^+) or ammonium forming fertilizers, such as urea, will cause a decrease in soil pH over time.
- Nitrate (NO_3^-) sources carrying a basic cation should be less acid-forming than NH_4^+ fertilizers.
- The presence of Ca, Mg, K, and Na in the fertilizer will slightly increase or cause no change in soil pH.
- Elemental sulfur, ammonium sulfate, and compounds such as iron or aluminum sulfates can reduce the soil pH.

Salt Damage

When inorganic fertilizers are applied to the soil, the concentration of soluble salts increases in the soil solution surrounding the zone of fertilizer application, particularly when there are high rates of evaporation and insufficient rainfall to leach the salts. A high concentration of soluble salts in the soil solution can have harmful effects on plants and germinating seeds. The salt index is a measure of the potential salt damage to the plant. Soluble salts can also originate from manure and minerals in the earth.

Problems associated with salt damage include:

1. If the concentration of salt in the soil solution is greater than the salt concentration in plant root cells, moisture availability will be restricted and water will not be absorbed by the plant.

2. The high concentration of salts will cause water to leave the plant by osmosis and enter the soil. Excessive water loss causes the protoplasm to shrink away from the cell walls (plasmolysis). As a result, the plant withers and exhibits symptoms similar to those caused by drought.

3. High concentrations of soluble salts may also result in elemental toxicities of sodium and chlorine.

Fertilizer Salt index

The fertilizer salt index was developed to classify fertilizers according to their potential to cause salt injury to plants. It is a measure of the salt concentration that a fertilizer material induces in the soil solution and it is measured by placing the material in the soil and determining the osmotic pressure of the soil solution. Sodium nitrate is the standard and has a salt index of 100. Other fertilizers are assigned a salt index value relative to 100, which describes the potential of the fertilizer to cause salt injury relative to the damage caused by an equal amount of sodium nitrate. A fertilizer with a salt index less than 100 has a lower potential to cause salt damage in comparison with a fertilizer with a salt index greater than 100.

$$Salt\ Index = \frac{Increase\ in\ osmotic\ pressure\ of\ soil\ solution\ caused\ by\ a\ fertilizer}{Osmotic\ pressure\ produced\ by\ the\ same\ weight\ of\ sodium\ nitrate}$$

Nitrogen and potassium salts have higher salt indices than phosphorus salts, hence they have more damaging effects on germination when placed close to the seed. Salt injury can be avoided by applying N and K fertilizers on the soil surface (broadcast application) or by placing them to the side and below the seed.

Soil Amendments

A soil amendment is a material added primarily to change or enhance the physical, chemical or biological characteristics of soil, rather than as plant food. Examples include liming materials such as lime (calcium carbonate), dolomite (calcium magnesium carbonate), calcium oxide, magnesium oxide, calcium hydroxide, magnesium hydroxide wood ash, and slags, which are used primarily to increase the pH of a soil and make it less acid. Liming materials should not be added with urea or ammonium fertilizers, as N will be lost to the atmosphere. Elemental sulfur is used to decrease pH and make it less alkaline. Gypsum (calcium sulfate) is used primarily as an amendment on soils with excess sodium (sodic soils), which causes a poor physical condition. The calcium in the gypsum replaces the sodium, resulting in a soil with improved structure. The sodium combines with the sulfate and is leached from the soil by irrigation.

These soil amendments also provide plant nutrients if they are needed. For example, limestone provides calcium and in some cases magnesium. Gypsum provides calcium and sulfur.

Animal Manure and Green Manure Crops

Animal manures are excellent sources of plant nutrients. Because manure is often derived from plant material, it is similar to organic matter and therefore contains essential nutrients. When manure decomposes in the soil, its nutrients are released and made available for plant uptake.

Manure not only serves as a source of plant nutrients but also adds organic matter to the soil, which improves its physical condition, water-holding capacity, cation exchange capacity, and other desirable properties. For maximum value, manure should be injected or worked into the soil soon after application (Fig. 8.14). A disadvantage of manure is that it often contains weed seeds.

The nutrient and moisture content of manure is quite variable, depending principally on the types of feed utilized by the animals and how the manure is handled before it is applied. It should be applied as often as convenient to do so, but in most cases, manure has to be stored and applied later. Alternate wetting and drying in a pile results in the release of ammonia gas into the atmosphere as the manure dries. When rewetted by rain, nitrates leach out and may present a danger if allowed to run off into a water system being used by humans or animals. Many states have laws that restrict the runoff from feedlots and manure piles, and require owners to provide holding ponds or lagoons (Fig. 8.15). If a manure pile dries without subsequent rewetting (in dry areas), loss of ammonia is minimized.

The average nutrient value of manure from beef feedlots in Texas is shown in Table 8.5. The nutrient content of manure is also quite variable; thus, only averages are

8.14   Animal manure improves soil structure as well as supplying nutrients.

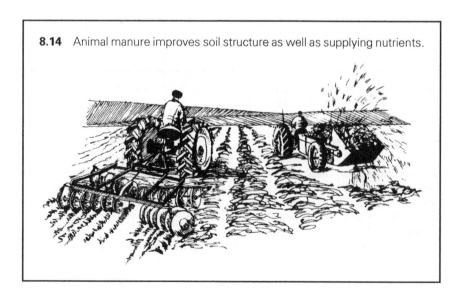

8.15 Lagoons provide storage and maintain the nutrient value of manure.

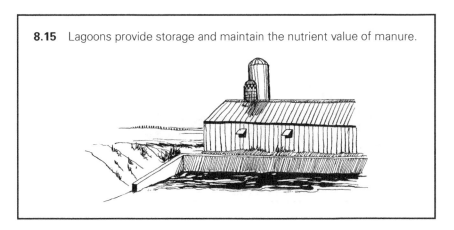

Table 8.5 Average content of essential elements in beef feedlot manure (based on 30 samples from Texas High Plains feedlots, figured at 30% moisture content) (Eash, Neal S., Cary J. Green, Aga Razvi, and William F. Bennett, eds. *Soil Science Simplified.* 5th ed. Ames, Iowa: Wiley-Blackwell, 2008. Copyright © 2008, John Wiley & Sons, Inc.)

Nutrients	Content	Pounds/ton (kg/metric ton) of manure
	(%)	
Nitrogen (N)	1.6	32 (16)
Phosphorus (P_2O_5)	1.3	26 (13)
Potassium (K_2O)	2.2	44 (22)
Calcium (Ca)	0.7	14 (7)
Magnesium (Mg)	0.2	4 (2)
	(*ppm*)[a]	
Iron (Fe)	1525	3.00 (1.50)
Zinc (Zn)	100	0.20 (0.10)
Manganese (Mn)	105	0.21 (0.105)
Copper (Cu)	7	0.02 (0.01)
Boron (B)	15	0.03 (0.015)

[a] ppm = parts per million or 1/10,000 of a percent.

presented in the table. The values of moisture content ranged from a low of 8% to a high of 62%, with an average of 33%. One ton (900 kg) of manure with a moisture content of 33% and average nutrient percentages would be equivalent to 400 lb (180 kg) of an 8-6-11 grade fertilizer.

When animals are confined in barns most of the time, their manure is commonly stored for 6 months or more in lagoons or large tanks and the moisture content is more likely to be around 90%. It is injected into the soil as a slurry from tanks on wheels (Fig. 8.16). This is typical for dairy farms. Under these conditions, the nutrient content is likely to be about one-third of that shown in Table 8.5.

The N/P ratio in animal wastes is typically 1:1 to 2:1 (N/P_2O_5), but most plants require 3:1 to 5:1 (N/P_2O_5). Therefore, animal waste is relatively high in P. If manures are applied to meet the N needs of plants, excessive amounts of P often result. Since P is relatively

8.16 A tractor-powered mobile tank and pump unit for injecting liquefied manure into the soil.

8.17 Crops can be plowed under as green manure to provide organic matter.

immobile in the soil, any P not taken up by plants will accumulate in soil to levels far in excess of amounts needed for optimal crop growth. The excess P in the soil can be lost via leaching, erosion and runoff and can cause nutrient enrichment in ground and surface water sources and cause eutrophication. A more environmentally approach is to apply manure to meet P needs (based on soil testing). This will result in lower N application, but the remaining N needs can be met with fertilizers. This approach requires more land since smaller amounts of manure will be utilized on each field.

The economic value of manure is also variable, depending on the nutrient percentages. In most cases, approximately one-half of the nutrients are released and available the first year. The rate of decomposition, however, varies widely. On this basis, if the analysis is known, a value can be placed on manure. Because manure has relatively low nutrient percentages, the volume that must be handled is relatively high if sufficient plant nutrients are to be applied. Consequently, manure is normally used fairly close to the farm or feedlot where it is produced. Manure application rates are generally 10–15 tons per acre (22–34 Mg/ha [megagram/ hectare]). Manure should normally be applied on an "as-is" basis for crops that have a significant nitrogen requirement.

Crops plowed under soil (Fig. 8.17) to improve fertility and physical condition of the soil are called green manure. The best crops for this purpose are legumes such as alfalfa and clover because they are high in nitrogen content; however, nonlegume crops such as wheat or sudan grass can also be used.

Fertilizer Placement Methods

The major function of correct fertilizer placement is to enhance nutrient availability and plant uptake. Correct fertilizer placement encourages maximum crop yields because it often improves the efficiency of nutrient uptake. This leaves less nutrients in the soil, thereby protecting both surface and groundwater quality.

Numerous placement methods are available but most generally involve surface to sub-surface applications before, at or after planting. Prior to planting, fertilizers can be broadcast on the surface or incorporated into the soil, applied as a band on the surface or applied as a subsurface band. At planting, fertilizers can be banded with the seed, below the seed or below and to the side of the seed. After planting, application is usually limited to N as a top-dress or a side-dress.

There is no single best method and several factors need to be considered in making fertilizer placement decisions. These include soil conditions, soil test level, soil P buffering capacity, crop type, salt effect of the fertilizer, convenience to the grower, and equipment availability.

Broadcast Methods. A broadcast involves the application of nutrients uniformly on the soil surface and may or may not be incorporated. A majority of fertilizer used in the USA is applied using broadcast methods. The popularity of broadcast pre-plant applications of solid fertilizers has increased in recent years due to the desire to: reduce fertilizer injury to plants, cut down on the amount of time spent handling fertilizers and the development of bulk blends that enable the application of a large quantity of material at one time.

Surface Broadcast. Surface broadcast is a method by which fertilizer is applied uniformly over the surface of an entire field (Fig. 8.18). It is best suited for high-speed operations and heavy application rates. High capacity fertilizer spreaders are often used to spin dry fertilizer or spray liquid fertilizer on the soil surface or on a growing crop. A broadcast application is referred to as a top-dress application when applied at planting, or when applied to the standing crop (after emergence of crop). Care must be taken in top dressing to ensure that fertilizer is not applied when the leaves are wet or it may burn or scorch the leaves. Broadcasting is the only option for applying N and K fertilizers to existing stands of perennial forage crops such cool season grasses and alfalfa.

Advantages:

1. It is fast, easy and economical (saves labor and time during planting).
2. It results in relatively uniform distribution of fertilizer.
3. It requires relatively inexpensive application equipment.
4. It can reduce work load at critical times during the year since there are several opportunities for application.
5. Large amounts of fertilizer can be applied without the danger of injuring the plant.
6. It provides a practical way to apply maintenance fertilizer, especially in forage crops and no-till systems.

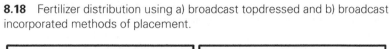

8.18 Fertilizer distribution using a) broadcast topdressed and b) broadcast incorporated methods of placement.

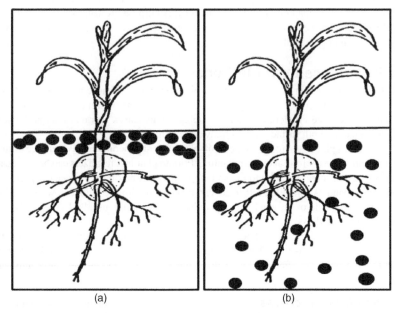

(a)   (b)

Disadvantages. Broadcasting has low efficiency of nutrient use because:

1. It enhances N immobilization and losses of N via volatilization in high surface residue systems such as no-till systems.

2. N and P losses by erosion as well as N losses by denitrification are higher compared with placement in the soil.

3. It leaves more fertilizer available to weeds.

4. It requires rainfall or irrigation to move mobile nutrients such as NO_3-N and SO_4-S into the root zone. If the soil dries out, these nutrients become unavailable.

5. It leaves immobile nutrients (P, K, and some micronutrients) on the soil surface, making them unavailable to the plant root system.

6. It does not increase the overall fertility of the soil as it does not store fertilizer deep in the profile for later season plant nutrition.

7. Uniformity of application can be poor for low rates of fertilizers.

Broadcast Incorporated. A broadcast application is referred to as a broadcast-incorporated when the broadcasted fertilizer is immediately incorporated into the soil by tillage, resulting in fertilizer being mixed throughout the root zone (Fig. 8.18). This improves on the efficiency of surface application. Plowing creates a nutrient-rich zone a few inches below soil surface where young plant roots can absorb it, thereby improving nutrient availability.

Advantages:

1. Reduces volatilization and erosion losses compared to broadcast placement.
2. Increases uptake of immobile nutrients (i.e., P and K) by increasing the probability of contact between plant roots and fertilized soil.
3. Improved plant uptake leaves fertilizer less available for weeds.
4. Distribution of nutrients throughout the plow layer encourages deeper rooting and exploration of the soil for water nutrients.
5. Incorporation promotes rapid nitrification of NH_4^+ to NO_3^+.
6. Incorporation with plow mold-board and plow chisel stores fertilizer deep in the profile for later season plant nutrition.
7. Reduces the chance of salt injury to seedlings.
8. Increases the overall fertility of the soil.
9. Maximizes contact between soil and fertilizer thereby creating more potential sites for adsorption and subsequent retention of K and Mg in low CEC (sandy) soils.

Disadvantages:

1. Requires more energy to incorporate the fertilizer.
2. Results in potentially higher N and S leaching losses than with surface placement.
3. Stimulates some weed growth.
4. Some tillage implements, such as a mold board plow, may distribute the majority of the fertilizer possibly too deeply for the roots of young seedlings.
5. Potential for leaching losses of N and sulfur (S) are higher than with surface placement, particularly in wet years.
6. Maximizes contact between the soil and fertilizer, thereby increasing the opportunity for tie-up/fixation and reduced availability of P and K.
7. Some implements, such as a chisel, incorporate fertilizer only 2–3 in. (5–7.5 cm) into the soil surface. This leaves nutrients inaccessible to roots when the surface soil dries out.

Nutrients that are generally broadcast include nitrogen, sulfur, calcium, and magnesium (as liming materials), copper, manganese, zinc (but banding is more efficient for all these micronutrients), and boron. Potassium may be applied by either broadcast or band methods. However, broadcasting is preferred in sandy soils that have a low CEC (less than 3 meq/100g) while banding is preferred in soils containing a lot of vermiculite and/or illite where K fixation occurs.

Nutrients that are generally not broadcast include iron (generally applied as a foliar spray) and molybdenum (generally applied as a seed treatment). Phosphorus is broadcast only in soils with moderate to high levels of P (at soil pH values of 5.5–6.5).

Pop-Up Application. Pop-up fertilizer applications refer to placing small amounts of nutrients in direct contact with the seed (Fig. 8.19) to enhance the availability of nutrients to young plants and enhance early seedling vigor. It is also known as starter fertilizer. Nitrogen and P are usually pop-up fertilizer components. The upper limit of pop-up fertilizer that can be used is determined by the relative salt tolerance of the plant. Large amounts of N, K, and S fertilizers cannot be used in pop-up applications due to their high salt contents. Boron cannot be used in a pop-up application as high concentrations of B are toxic to plants.

8.19 Pop-up or direct seed contact method of fertilizer placement.

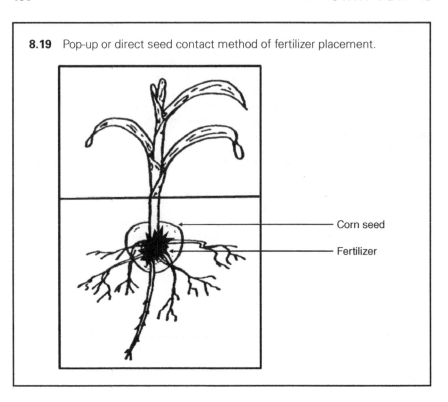

Corn seed

Fertilizer

Solid fertilizers used in pop-up applications should be highly soluble in water. They should also have a low salt index, a high nutrient content and minimal content of N materials that produce ammonia.

Advantages:

1. Both liquid and dry fertilizers can be used.
2. Promotes early plant growth and enhances seedling vigor. This early stimulation of crop growth is often termed "pop-up effect."
3. Decreases loss of nutrients by erosion compared with surface placement.
4. Positions fertilizer where root systems of seedlings can more readily use nutrients.
5. Positions fertilizer so it is more available to the crops than to the weeds.
6. Placing P with the seed increases the concentration of available P sufficiently to partially offset the detrimental effects of extremes in soil pH, compaction and low soil temperatures under conservation tillage systems.

Disadvantages:

1. Can cause seedling damage if too much fertilizer is applied.
2. Retro-fitting planters can be expensive.
3. Urea and diammonium phosphate cannot be used.

Total application rate must be kept below 10 lb (4.5 kg) of N + K_2O so as to prevent salt burn.

8.20 Band fertilizer placement method.

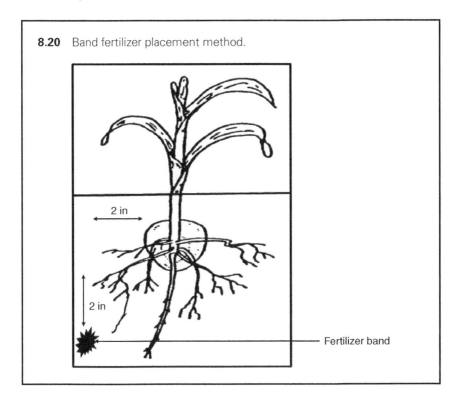

2 in

2 in

Fertilizer band

Band Application. Banding refers to placing nutrients below, above, on one side, or on both sides of the seed or seedlings so that developing roots will easily reach the nutrients (Fig. 8.20). Fluid or solid fertilizers can be used. All plant nutrients, with the exception of boron (B), can be successfully banded. High concentrations of B in soils are toxic to plants.

Banding can be done before or at the same time with planting/drilling, or after the crop is planted. A low rate of fertilizer is placed in close proximity (at least 2 in./(5 cm) to the side and 2 in. (5 cm) deeper than the seeds or plants). This provides the plants with a concentrated zone of nutrients while preventing salt damage and ammonia toxicity.

Advantages:

1. Requires less fertilizer per acre than broadcasting.
2. Jump-starts early plant growth by increasing P availability.
3. The positioning of fertilizers is such that nutrients are more available to the crops than to the weeds.
4. Decreases P and K fixation by limiting surface area of contact of fertilizer with the soil, thereby increasing their availability to plants.
5. Absorption as well as movement of P and N to plant roots is much slower at lower soil temperatures. Banding applications of N and P improve plant uptake at lower soil temperatures. This improves growth, thereby promoting winter hardiness.
6. Fertilizer application and planting operations can be done simultaneously.
7. Reduces nutrient losses due to soil erosion.

8. Similar to broadcast – incorporate, banding ammonium-based N fertilizers (e.g., ammonium nitrate, ammonium sulfate, and urea) below the soil surface reduces volatilization losses.

9. It slows NH_4^+ conversion to NO_3^- (nitrification), reducing the risk of leaching.

10. Enhances seedling growth thereby resulting in stronger seedlings that are less prone to suffer from pests and diseases.

11. Has high nutrient use efficiency.

Disadvantages:

1. There is risk of salt burn to plants.

2. There is increased NO_3^- and SO_4^- leaching losses compared with surface placement.

3. Slows planting if applied with a drill.

4. It is costly as it requires expensive equipment or equipment modification.

Side-Dress Application. A surface or subsurface banding treatment after the crop is planted is referred to as a side-dress application. In this case, fertilizer is applied between rows of growing plants to supply N during periods of rapid growth as this is when the crop needs it most. The most common use of side-dress application on farms is the application of N fertilizer between the rows of a growing corn when the plants are 12–24 in. (30–60 cm) tall. The amount of fertilizer N to be applied is determined based on the results of a Pre-Side-dress Nitrate Test (PSNT), which assesses the likelihood of a yield response to the addition of side-dress N. Side-dress application is not recommended for immobile nutrients (P and K) because most crops need these early in the growing season.

Advantages:

1. Provides a valuable opportunity to apply the recommended nitrogen throughout the season in smaller amounts (split application), rather than applying all the nitrogen in a single application.

2. Applications can be made whenever the equipment can be operated without damage to the crop. This allows a grower more flexibility in application time.

3. Split or multiple applications provides N when the plant needs it most, resulting in high nutrient use efficiency.

4. Provides crops with additional nutrients if applied during the growing season.

Disadvantages:

1. Timing often falls during the wet and busy season; slow process.

2. Subsurface side-dress applications with a knife too close to the plant can cause damage to roots by pruning or by fertilizer toxicity.

Point Fertilization. This consists of opening a hole in the soil with a stick or hoe, and placing a quantity of fertilizer into the soil near the crop. This system is commonly used in the production of perennial shrub or tree crops. It is also used in many developing countries, where competition from weeds is minimized by placing small quantities of fertilizer in a hole dug near a hill of sorghum or corn.

Fertigation. Fertigation is the application of dissolved fertilizers and chemicals to the soil through an irrigation system, which applies both water and nutrients to plants. The principal nutrients applied by fertigation are N and S. Fertigation of P is less common due to concerns over precipitation of P in waters high in Ca and Mg.

Advantages:

1. Nutrients can be applied close to the time of peak crop demand, thus preventing luxury consumption of nutrients.
2. May reduce losses of nitrogen due to leaching and denitrification.
3. Provides an opportunity to split nitrogen recommendation into several applications.
4. Results in reduced operation costs through elimination of one or more field operations.
5. Nutrients may be applied continually throughout the crop growing period. This enables a high degree of flexibility in nutrient management.
6. Can be used to correct midseason nutrient deficiencies.
7. Combining fertilization and irrigation into a single field operation saves on time and labor.
8. High nutrient use efficiency.

Disadvantages:

1. Irrigation equipment needed (injection pump, etc.).
2. There is risk of uneven distribution of nutrients with low rates of fertilization, in windy situations and under row irrigation.
3. Requires a well-managed and equipped irrigation system for uniform, maximum efficiency.
4. Nutrients may be leached beyond the root zone.

Foliar Application. This is the application of a small amount of soluble fertilizer or mineral through direct spraying onto the aerial portion of plants. This is a common way to apply micronutrients since micronutrients are required in much smaller quantities than macronutrients.

Advantages:

1. Foliar fertilizers supply plant cells with nutrients more rapidly than the soil.
2. Can provide a quick way to correct nutrient deficiencies.
3. Very effective for micronutrients as these are required in small amounts.
4. It is the most effective means of fertilizer application in situations where the problem of nutrient fixation by the soil occurs.

Disadvantages:

1. It is expensive.
2. It is limited to small and/or repeated applications.
3. Has limited capacity to supply macronutrients because of salt hazards.
4. The response is usually temporary, hence repeated applications are required.

Injection. Injection is used to place liquid or gaseous fertilizer below the soil near plant roots.

Advantages:

1. Reduced volatilization losses through precise application of nutrients.

Disadvantages:

1. Slow and expensive (requires specialized equipment).
2. Mobile nutrients may be lost via leaching

Timing of Fertilizer Application

Determining the appropriate time to apply fertilizers in the soil is just as important as choosing the correct amounts of plant nutrients and determining the proper zone in the soil in which to apply the fertilizer. Farmers sometimes apply fertilizer soon after the previous year's harvest. While this coincides with availability of equipment and labor, this may not be efficient agronomically as fertilizer applied too far in advance of crop demand may be lost, resulting in negative economic and environmental consequences. Optimal timing of fertilizer application ensures an adequate supply of nutrients during peak and critical crop demand periods. This maximizes nutrient recovery by the crop, thereby reducing the potential for loss of nutrients from the system.

Proper timing of fertilizer applications in the field aims to:

1. Provide a sufficient amount of the nutrient when the plant needs it.
2. Avoid excess availability of nutrients, especially of N, before or after the principal periods of plant uptake (especially during environmentally sensitive periods of groundwater recharge).
3. Make nutrients available when they will strengthen, not weaken, plants.
4. Enable field operations to be conducted when it is feasible.

Some basic factors to be considered in fertilizer timing decisions.

1. *Nutrient element*: The greater the potential for loss of a given nutrient from the soil, the greater the importance of timing of application. Proper timing is critical with nitrogen fertilizer, as N is susceptible to loss from the soil through several pathways including leaching, denitrification, volatilization, and runoff/erosion (see Chapter 4). For example, fall application of N for spring-planted crops, such as corn, should not be practiced in humid areas because of high risk of loss. To slow nitrification in the fall and avoid increased nitrate leaching and/or denitrification, fall application of N should only be practiced in late fall after the soil temperature drops to below 50°F (10°C) and expected to continue cooling, even in drier areas with low risk of loss. Time of application is less critical with P and K and fall application is generally considered a reasonable practice as the risk of runoff is small in that season. However, phosphorus application is most efficient when made at or as close to planting time as possible, especially in low P soil and/or soils with high capacity to convert soluble P into less available forms. Application timing is also an important consideration with potassium in sandy, low cation exchange capacity soils in high-rainfall areas, because of the high potential for leaching in these environments.

2. *The N form*: The N form in a fertilizer product can affect the potential for loss and optimal timing. In southern US, the temperatures favor nitrification during a greater portion of the year, hence ammoniacal N applied before planting would be converted to nitrate and lost via leaching. Fertilizer materials containing significant portions of nitrate are not suggested for fall application, because of leaching and gaseous losses. Instead, spring pre-plant and/or side-dress applications are preferable as they typically provide a lower risk of loss and greater profitability. And, even with spring application, materials containing significant portions as nitrate are more susceptible for loss with early pre-plant application.

3. *Crop and plant nutrient uptake pattern*: Crop nutrient demand is not consistent throughout the growing season but it is characterized by an initial stage of slow uptake, followed by a phase of rapid uptake, followed by a period of declining uptake as the crop matures. Therefore fertilizer applications, particularly N, can be timed and targeted at specific growth stages to increase N uptake (crop yield and/or quality) and to reduce the potential for loss to the environment. Timing of fertilizer application can also be targeted to correct specific nutrient deficiencies during the growing period.

4. *Soil characteristics and environment*: Certain site and soil characteristics influence the potential for nutrient loss, nutrient retention, and supply capacity and are important considerations in decisions of fertilizer timing. The characteristics include slope, soil texture, temperature and drainage. The greater the soil's capacity to retain and supply and provide a crop-available nutrient throughout the growing season, the less the need for a critical timing emphasis for that nutrient. For example P and K fertilizers can be applied once on most soils in the Corn Belt to supply crop needs. In contrast, P fertilizer products need to be banded at or near planting time in soils with very high P fixation capacity such as the highly weathered soils in the southern USA and the calcareous soils of the West. Proper timing of N fertilizers is important on sandy-textured soils in areas receiving high rainfall during winter and spring. Nitrogen fertilizer applications on these sandy soils are usually split into two or three applications to increase N uptake and reduce the potential for loss to the environment. Conversely, there is little advantage to splitting application of N between fall and spring in arid environments with low loss potential, hence all N may be applied pre-plant.

Other factors affecting fertilizer application timing decisions include feasibly of conducting the operation, farm size, other field operations, availability of equipment and labor, and fertilizer distribution logistics.

Precision Farming

The computer age has led to many innovations in agriculture. One such innovation is precision farming. Because soils in most farm fields vary considerably in such properties as organic matter, topsoil thickness, texture, structure, and plant nutrient content, it is inefficient for an entire field to receive the same amount of fertilizer when the crop yield potential and soil nutrient level vary from one area of the field to another. It is claimed that fertilizer costs are reduced, the environment is better protected, and crop yields are higher if the proper amount of fertilizer is more precisely applied to each of the various management zones in the field. It can be demonstrated that the variations in soil properties are often more detailed than the soil maps discussed in Chapter 11.

The first step in using precision farming techniques involves sampling the soil by using an ATV (all terrain vehicle) in a grid pattern with each sampling point commonly representing 2.5 acres (1 ha) (see Fig. 8.5). Management zones can also be used. The position of each sample site is monitored and programmed into the computer via the GPS (global positioning system), which relies on a set of satellites about 200 miles (320 km) above the earth. The ATV is equipped with a receiver that collects signals from at least three satellites for accurate positioning by triangulation (Fig. 8.17). With an enhanced system, the sampling sites usually can be pinpointed to within 1 m. Each sampling location appears as a point on a computer screen.

The second step is to analyze the samples for nutrient content and other soil characteristics (such as pH and organic matter). From these results, a pattern of the fertilizer requirements throughout the field can be established and stored on a computer disk.

The third step requires a fertilizer applicator with separate bins for individual fertilizer, each equipped with augers to transfer their contents to the spreading apparatus. The augers are regulated by a computer in the cab of the applicator whose position has been determined by the GPS. In this way, there can be a continuous adjustment in the rate and mixture of fertilizer applied as the truck goes back and forth across the field.

In the fourth step, the combine that harvests the crop is equipped with a device that continuously monitors the yield of the grain harvested from all parts of the field in relation to a previously programmed grid sequence described in the first step. The yield map is used to set the yield goal, which is a factor used to determine the amount of nutrient to apply.

Farmers may contract for the precision farming techniques described previously and many are currently doing so. The extent to which precision farming becomes a common practice depends on how farmers judge its economic value.

Organic Farming/Gardening

Organic farming or gardening involves producing crops without applying commercial fertilizer or chemical pesticides. Organic farming usually includes the slow release of a naturally balanced supply of nutrients from decaying organic matter (Fig. 8.18), such as crop residues and animal manure. Crops of high quality and quantity may be grown by organic methods. In parts of the world where commercial fertilizers are not available, farmers must rely on decaying organic matter for supplying crops with nutrients. Enough food could not be provided for the world's population if the plant nutrients for food crops were to be supplied solely from organic sources. The recycling of carbon and nutrients from plant and animal manures is, however, an important benefit of organic production techniques. Some consumers prefer to eat food grown without chemicals, so they are willing to pay higher prices for crops that are grown using organic methods.

Nutrients from the soil are absorbed by plants mainly in the form of ions (charged particles). Whether these ions come from a weathering mineral, decaying humus, or a chemical fertilizer is of no consequence to the plants. But the nutrients should be in a properly balanced proportion.

Composting

Composting is a planned and managed process to foster the aerobic decomposition of organic matter. Unlike anaerobic decomposition, compositing doesn't produce odors and seeping noxious liquids.

Composting is often used for a source of organic matter for gardeners (Fig. 8.19). The basic ingredients of a compost heap are organic residue, soil, moisture (50–70%), nitrogen fertilizer, and some lime to counteract acidity associated with decomposition. The carbon-to-nitrogen ratio of the material must be close to 30:1. The C/N ratio can be lowered by adding nitrogen fertilizer or by mixing high carbonaceous materials such as straw with low carbon materials such as livestock manure, green grass clippings, and legume hay. Air (oxygen) must be able to diffuse through the pile to the decomposing residue and prevent anaerobic decomposition from taking over. The heap is allowed to decompose (rot) for several months and is then mixed into the soil. The rotting process is carried out by microbial action that is greatly hastened in the presence of an adequate amount of nitrogen. Other materials such as rock phosphate powder, wood ashes, and mixed fertilizers may be added for improved nutrient balance of the final product.

Composting is also becoming more popular for organic farming. A typical system has large rows of organic material (leaves, sawdust, or manure) that is amended with an organic source of nitrogen, and other nutrients if necessary. A machine mixes the material in the rows until the compost is stable (partially decomposed). The heating of the compost, as some of the organic material decays, has an added benefit of killing most of the pathogenic or disease-causing microbes. Finished compost can be bagged for sale in stores or loaded onto trucks or spreaders for application to fields.

Composting is a good way to turn most organic waste products into a reliable and valuable plant nutrient source. However, certain wastes are problematic and they should be avoided. For instance; meat scraps attract rodents and other pests and produce noxious odors as well, cat droppings carry microbes that may be harmful to infants and pregnant women, plastics and glass are non-biodegradable, while plywood may be laden with heavy metals.

Biosolids

Biosolids or sewage sludge, are nutrient-rich organic materials resulting from the processing of domestic sewage in a treatment facility. Biosolids contain nutrients such as nitrogen, phosphorus, and potassium and trace elements such as calcium, copper, iron, magnesium, manganese, sulfur, and zinc. Some biosolids might also be lime-stabilized, resulting in a pH increase when applied to the soil. In addition, the organic matter in biosolids has several benefits (see Chapter 4). Therefore, when used in accordance with existing federal guidelines and regulations, biosolids can be applied as part of a crop nutrient management plan to produce crops for human consumption with negligible risk to both the consumer and the environment. When used to fertilize crops, the nutrient content of biosolids should be tested and their contribution to the total crop nutrient needs should be accounted for. Furthermore, recommendations for the appropriate application rate, method, and timing should be followed.

Since biosolids are regulated byproducts of wastewater treatment, they pose various health and safety risks that must be addressed when they are applied to the soil. These include: presence of disease-causing organisms and heavy metals, odor, and insect problems associated with the application of raw, unstabilized wastes; and surface and ground water contamination by nitrogen, phosphorus, and pathogens.

CHAPTER 9

Soil Management

Proper soil management is an important part of any operation that involves crop production or natural resource management. The goal of soil management for crop production is to establish and maintain the correct combination of all soil factors necessary to optimize and maintain production efficiency. Effective soil management will ensure that food and fiber production are maximized and sustained over the years while leaving the soil in a productive state.

In agriculture, soil management is closely related to crop management. It includes (1) maintaining the soil in a good physical condition; (2) keeping the chemical characteristics of the soil in the proper balance, such as maintaining soil fertility and the correct pH; and (3) influencing the biological or organic portion of the soil so that maximum benefits result.

Physical Condition

Soil with good physical condition is important to plant growth. The physical condition of soil as it relates to ease of tillage, quality of the seedbed, and resistance to seedling emergence or root growth is referred to as soil tilth. Good tilth helps the water infiltration rate, water-holding capacity, soil–air interchange, and root development. It also aids in reducing erosion. Soil tilth can be maintained by continuing to return plant residues and organic materials to the soil and by using tillage practices that do not compact soil. Physical condition deteriorates if the soil is compacted by tilling when it is too wet or by using heavy machinery too often.

Residue Management

Crop residues influence the soil's physical condition. A soil that is loose and friable is generally considered to have good tilth. Managing the residues on the soil surface is important to maintaining the tilth. Often the old crop residues are incorporated (buried) into the soil during tillage. The first residue management step is deciding if and when

Soil Science Simplified, Sixth Edition. Neal S. Eash, Thomas J. Sauer, Deb O'Dell, and Evah Odoi.
© 2016 John Wiley & Sons, Inc. Published 2016 by John Wiley & Sons, Inc.

9.1 Grasses are low in nitrogen at maturity and are more slowly decomposed than legumes, which contain much more nitrogen.

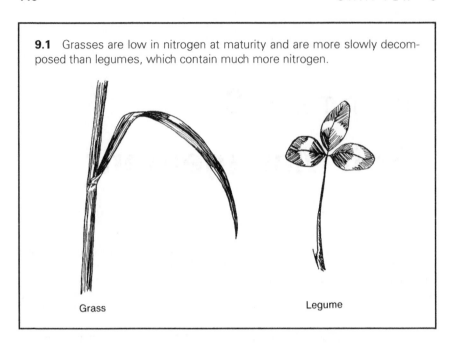

Grass   Legume

crop residue incorporation is needed. The time to incorporate residue will depend on the cropping system. Options for residue management range from complete incorporation of residue—often referred to as clean till—to using no-till where all the old crop residues are left on the soil surface.

Residue may be incorporated any time from immediately after harvest to just before planting of the next crop. If the field is to be planted the following year, it is sometimes desirable to incorporate residue soon after harvest to start the decomposition process. This helps to ensure that a smooth seedbed can be prepared and that the soil is free of excessive residue for planting. If a crop is not going to be planted the following spring, the residue can be left on the soil surface to act as a cover. This would greatly reduce erosion by both wind and water and conserve soil water.

Where wind or water erosion is a problem, partial incorporation of residue into the soil may be desirable, with a portion left on the surface (stubble mulching). Some farmers never incorporate plant residues into the soil and plant the subsequent crops with the residue still on the surface (minimum tillage).

As discussed in Chapter 4, bacteria decompose (break down or change) crop residue after it has been incorporated into the soil. Bacteria break down corn, wheat, and grain sorghum (all in the grass family) residues quite slowly, whereas alfalfa, soybean, and clover (all in the legume family) residues decompose more rapidly. The principal reason for the difference in decomposition between the two groups is the amount of nitrogen (in protein form) in the residue (Fig. 9.1). Legumes are high in nitrogen, whereas grasses are low in nitrogen. If it is ever desirable to speed up the decomposition or breakdown of a low-nitrogen residue, adding nitrogen fertilizer should help.

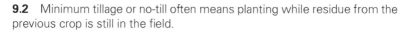

9.2 Minimum tillage or no-till often means planting while residue from the previous crop is still in the field.

Tillage Practices

Tillage practices used to manage soil vary widely. The type of crop grown, type of soil, erosion hazards, the use (or not) of irrigation, and cost of the tillage practice are all considered in determining which practices to use.

Tillage of soil may vary from farms on which there is no tillage (except planting)—called no-till or minimum till systems—to farms where the soil may be tilled 8–10 times a year. Because of excessive erosion and high costs, the trend for most producers is to reduce the number of tillage operations. Many farmers now use a reduced or conservation tillage system, where some crop residue is left on the surface and the soil is only moderately disturbed. These systems can be thought of as a compromise between the extremes of tillage, moldboard plowing, and no-till. Reduced tillage was initially adopted mostly by farmers in the higher rainfall areas of the eastern and midwestern parts of the United States.

In a no-till system, all the crop residue is left on the surface and seeds are planted with as little disturbance of the residue and soil as possible. An example of no-till planting is illustrated in Figure 9.2. In addition to cost savings, minimum tillage also decreases soil erosion and normally increases the water infiltration rate. Minimum tillage may require a modification of certain practices, including planting, fertilization, and application of pesticides. Fertilization rates, particularly for nitrogen, may need to be increased because soil temperatures tend to be lower as a result of the surface residues. Surface residue could also harbor insects, thereby resulting in an increased need for insect control.

Soil erosion by water and wind has been a problem since soils have been tilled. Conservation tillage practices were developed to reduce the loss of valuable topsoil by erosion; thus, they often involve leaving a portion of crop residues on the surface of the soil. By definition, conservation tillage is in effect when 30% of the soil surface is covered with crop residue. Conservation tillage has been practiced for many years in the areas of limited rainfall and potential wind erosion in the western United States. Crop residue is left on the surface of the soil not only to decrease erosion but also to conserve water (stubble mulching).

9.3 Disking incorporates crop residue to a shallow depth, a moldboard plow covers the residue, and a chisel plow goes deep but leaves no residue on the surface.

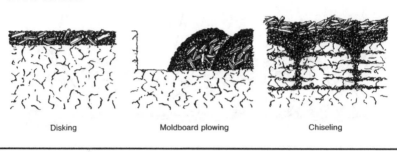

Disking   Moldboard plowing   Chiseling

Reasons for Tillage

Tillage refers to the moving, turning, or stirring of the soil. The soil is tilled to accomplish a number of things:

1. *Incorporate residue.* Incorporation of some crop residue into the soil hastens its decomposition. Without the decay of crop residues, it could become difficult to prepare a good seedbed and plant the seed. Different degrees of residue incorporation are illustrated in Figure 9.3; the implements employed are discussed in the next section.

2. *Improve physical condition.* Too often soils are worked or crops are harvested when the soil is too wet, causing the soil to compact or form a plow pan (Fig. 9.4). Some soils have naturally compact layers; deep plowing or chiseling can be done to break up the compacted soil or plow pan.

3. *Reduce erosion by wind or water.* Some tillage practices are used to reduce erosion (see Chapter 10). They include plowing on the contour, terracing, furrow diking, stubble mulching, and creating ridges to roughen the soil surface.

4. *Prepare the soil for planting.* This may consist of cultivating beds in rows in which to plant seed. Or it may involve a light disking or harrowing to break up a surface crust and at the same time destroy small weeds.

5. *Incorporate pesticides, fertilizers, and animal manures.* Some pesticides and fertilizers may be left on the soil surface and still be effective. Most pesticides, however, are incorporated either with a light disking or by using a rotary hoe. Fertilizers and manures are often incorporated into the soil by either disking or plowing.

6. *Control pests, including weeds, insects, and diseases.* Even though many types of weeds are controlled by chemicals, tillage practices are also used. A light disking before (or during) planting may be used to kill early-emerging weeds. Deep moldboard plowing is occasionally used to turn up roots of hard-to-kill weeds such as Johnsongrass so that they will be killed by freezing. Shallow tillage may be used on fallow land to control weeds. Tillage is also used to incorporate residues or host plants that might harbor insects and diseases.

7. *Increase water infiltration.* In areas where moisture is often limiting or where irrigation is practiced, certain tillage methods are used so that water, particularly rainfall, can move into the soil more rapidly or stay on the surface until it does so. Tillage that breaks up a surface crust can increase the infiltration rate. Forming rows on the contour and terracing as well as furrow diking (see Chapter 10) helps soil hold water longer.

9.4 Plow pans can form at the depth of tillage and inhibit root penetration because of their increased density. Chiseling or periodic deep plowing can prevent this effect.

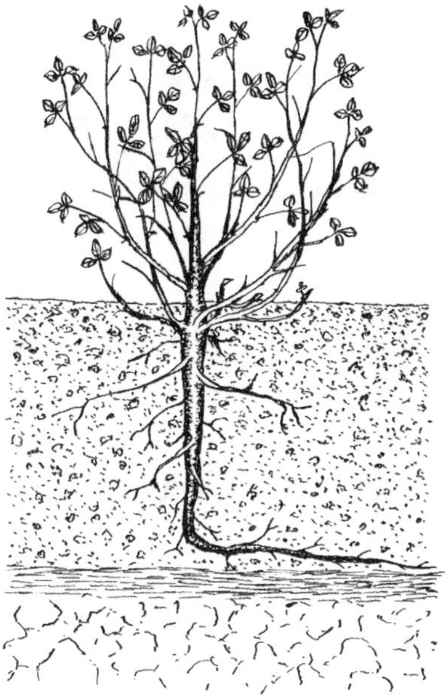

Tillage zone

Dense plow pan

Subsoil

Tillage Implements

Several types of farm implements are used to accomplish the tasks listed above. The degree of soil disturbance varies by the type of implement, how deep it is placed in the soil, and how fast it passes through the soil. Primary tillage implements refer to those that disturb the soil a great deal and incorporate significant amounts of crop residue (moldboard plow, chisel plow, and disk), while secondary tillage implements (disk, field cultivator, and harrow) generally till the soil to a shallower depth and often follow primary tillage operations.

A moldboard plow (Fig. 9.5A) is used to lift the soil and completely or partially turn it over. This can be done at any depth, but in most areas, it is 6-10 in. (15-25 cm). Development of a moldboard plow with steel shares (the "share" is at the base of the plow; it cuts and lifts the soil onto the "moldboard" which inverts the soil) was one of the key factors that enabled development of the Midwest and Great Plains of the United States as major crop-producing areas in the nineteenth century.

Moldboard plowing is often done to incorporate residue and break up dense soil. When followed by some secondary tillage, moldboard plowing helps create a fine seedbed for planting. Moldboard plowing incorporates most of the crop residues, which may be good for residue decomposition and seedbed preparation but can also leave the soil vulnerable to erosion. It was once the most common type of primary tillage but its use has decreased significantly due to the high cost of fuel to pull the plow and concern regarding excessive erosion.

9.5   Implements used for farming.

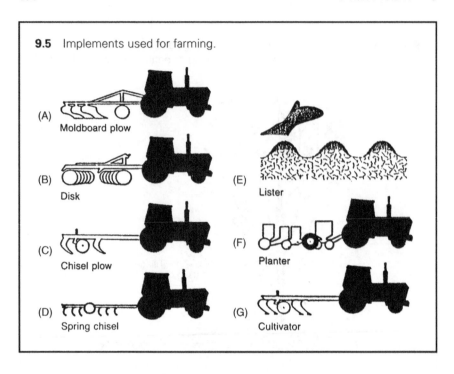

(A)   Moldboard plow

(B)   Disk

(C)   Chisel plow

(D)   Spring chisel

(E)   Lister

(F)   Planter

(G)   Cultivator

Deep plowing to a depth of 18–36 in. (46–92 cm) is done in certain areas. This is usually used on a soil that has a sandy surface and a clay subsoil, with clay being brought to the surface to mix with the sandy portion. In some regions, it is also used to incorporate underlying mineral soil into a surface layer of thick muck (high organic material).

A disk (Fig. 9.5B) can be used for both primary and secondary tillage by varying the depth of the implement. It is often used to incorporate materials into the surface of the soil, usually the top 4–6 in. (10–15 cm). This often includes chemicals for weed control, fertilizers or other soil amendments, or crop residues. A light disking is also used to control weeds. Disking or shredding of crop residues with an implement called a flail chopper sometimes precedes moldboard plowing to improve residue incorporation and prevent clogging of the plow. One disadvantage of disking is that it tends to compact the soil just below the depth being tilled.

A chisel plow is another type of implement used in primary tillage (Fig. 9.5C). It is an implement pulled through the soil usually at depths of 10–14 in. (25–35 cm) and used primarily to break up a hardpan or plow sole (a dense, compacted layer of soil usually caused by farm implements such as the moldboard plow). Some chisel plows are made so that the chisel vibrates, causing the soil to shatter or loosen. Recently, several companies have produced implements that combine a disk, chisel plow, and harrow. These implements are quite popular because the depth, spacing, and size of the individual elements can be customized or adjusted to suit the farmer's needs and the field conditions. Such implements allow farmers to use a single implement for most of their tillage operations.

A spring chisel or Graham-Hoeme plow is a tillage instrument used in the low-rainfall areas of the United States (Fig. 9.5D). Its principal advantage is that the soil is tilled to control weeds (and partially incorporate residue) while minimally disturbing the surface soil and thereby reducing loss of soil water by evaporation.

A lister may also be used for tillage (Fig. 9.5E). It forms the soil into beds or rows 6–8 in. (15–20 cm) high where the seeds may be planted or the bed may be lowered 3–4 in. (7.5–10 cm) before planting.

A planter (Fig. 9.5F) and a cultivator (Fig. 9.5G) may also be used to till the soil. Even though the principal function of the planter is to place the seed in the soil, the furrow openers and other attachments are capable of cutting through crop residues so fertilizers and pesticides can be applied during planting. Cultivating was usually done in the past to control weeds that germinated after planting. However, modern herbicides have replaced much of the requirement for cultivation. Cultivation is still used in some areas when chemical weed control fails and to break up crusts to improve infiltration.

Chemical Characteristics

It is also desirable to keep the chemical characteristics of the soil in proper balance or condition. This includes maintaining soil pH in an optimal range, providing a sufficient and balanced supply of nutrients, preventing or alleviating saline–sodic soil conditions, and avoiding soil degradation from toxic pollutants.

Soil pH

Soil pH (see Chapter 5) is important in crop production and is an indicator of the acidity or alkalinity of a soil as well as an indicator of levels of certain nutrients and their availability. It also influences biological activities in the soil.

If a soil is too acid (below pH 5.0), phosphorus, iron, and certain other nutrients have limited availability, and levels of calcium, magnesium, and potassium will be low. If a soil is too alkaline (above pH 7.8), phosphorus, iron, zinc, manganese, and other micronutrients will have reduced availability but there will be an ample supply of calcium, magnesium, and potassium.

Soils in the eastern United States and in many of the high-rainfall areas worldwide are more likely to be acid. If soils in these areas are in crop production, the pH usually ranges from 5.0 to around 7.0. A soil test is the best method for determining whether pH needs to be adjusted. Most soils are not adversely affected by acidity if they are in the pH range of 6.0–7.0. If the soil pH is 6.0 or below, it may be necessary to raise the pH (make it less acid) by the use of limestone (calcium and/or magnesium carbonate). Limestone for application to soil (also called aglime) is relatively inexpensive because it can be quarried from abundant deposits (Fig. 9.6).

The amount of limestone needed per acre (hectare) depends on the cation exchange capacity (CEC) of the soil (see Chapter 5). The CEC is dependent on texture, type of clay mineral, and organic matter content. At a pH of 5.5, a soil with a high CEC might need 3 tons of limestone per acre (6.72 Mg/ha [mega-grams/hectare]) every 2–3 years, whereas a soil with a low CEC may need only 1 ton per acre (2.24 Mg/ha) to raise the pH to a desired level (Fig. 9.7). The application rates and amount of time between applications depend on the crops being grown, the amount of rainfall and leaching, and the CEC.

Limestone increases pH (decreases the acidity) by providing calcium, and in some cases magnesium, which in turn replaces hydrogen or other acidic ions on the exchange complex. With more calcium and magnesium, which are bases, and less hydrogen on the exchange complex, pH increases.

In the western United States and other limited rainfall areas of the world, soils tend to be neutral to alkaline. If the soil pH needs to be decreased (made less alkaline) and the

9.6   Agricultural lime is produced from limestone quarried from bedrock.

9.7   Many humid-region soils need regular applications of lime to combat acidity.

soil is not sodic (to be discussed later), elemental sulfur (S) may be applied. Using sulfur to decrease the pH of the entire soil mass would be quite costly; hence, the economic benefits usually would be less than the returns. Where high pH due to calcium is a problem, the common practice is to apply sulfur or an acid-forming sulfur product into a small band or limited soil area.

The sulfur forms an acidic microenvironment in which nutrient availability may be greatly increased. In this way, a small amount of sulfur at a low cost can be beneficial for one

9.8 An illustration of how profit from fertilizer is maximized. In this hypothetical example, the most profitable rate of fertilization is 100 pounds per acre (about 110 kg/ha). Note that the maximum yield does not correspond to the most efficient rate of application.

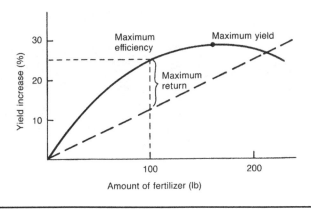

season in counteracting the undesirable effects of high pH. The amount of sulfur required to effect this change in a microenvironment may range from 20 to 100 lb of sulfur per acre (22–110 kg/ha) if properly applied. Sulfur products most commonly used for this purpose are prilled sulfur (80–90% elemental S), ammonium thiosulfate (26% S), and ammonium polysulfide (45% S).

Nutrient Supply

Keeping plant nutrients at adequate levels is important and proper nutrient balances need to be maintained to ensure that nutrient levels are adequate but not too high. Next to irrigation, this is probably the single most important soil management factor that can influence yield over which the producer has control. If too much of a nutrient is applied, not only is it an inefficient use of resources (Fig. 9.8) but it also may alter the balance of nutrients, adversely affect plant growth, and create a potential source of pollution. For a thorough discussion of nutrients and their application, see Chapter 8.

Saline and Sodic Soils

Another important management practice is to reduce the effect of saline and sodic soils on plant growth. A saline soil is one in which soluble salts have accumulated in sufficient quantity to adversely affect growth. A sodic soil is one that contains too much sodium, which adversely affects yields. Remediation of saline and sodic soils may involve a combination of soil management practices.

Saline and sodic conditions may occur naturally, but most arise when irrigation water is applied that is too high in salt and/or sodium. Areas in the United States where these situations occur are mainly in the Southwest, from Texas to California. It is a potential hazard in any irrigated area in the world.

Irrigation waters can be chemically analyzed to determine whether salt and/or sodium are high enough to create problems if used. Chemical tests are highly desirable for any new irrigation project or in areas where the salinity of water might tend to change.

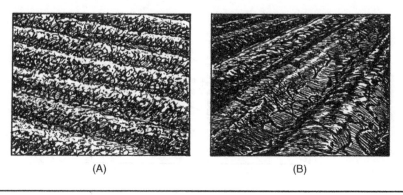

9.9 Saline soils (A) usually have "white caps" of salt in the tops of the beds. Growth of crops normally is spotted. Sodic soils (B) are usually dark colored (often called "black alkali") and are gummy and slick when wet and cracked with a powdery surface when dry.

(A)   (B)

Soils that become saline show irregular growth of crops in a field and usually have a whitish cast from salt accumulation, with the greatest amount being in the tops of the beds (Fig. 9.9A). Saline soils are often called "white alkali" because they are light in color and have an alkaline soil reaction.

If saline conditions develop, leaching the soil with water with less salt is a common practice. The downward movement of water in the soil that occurs during leaching carries the salts below the root zone, where they cause less of a problem.

If a soil becomes sodic, it is highly dispersed and in very poor physical condition (see Soil Aggregation in Chapter 5). The soil feels slick and gummy when wet. When dry, the soil is dark, appears to be highly dispersed, and has cracks. Sodic soils are called "black alkali" because of their dark color and alkaline soil reaction (Fig. 9.9B). Sodic soils usually range in pH from 8.5 to 10.0.

To correct a sodic soil condition, a calcium-containing compound— specifically gypsum, which is calcium sulfate ($CaSO_4 \cdot 2H_2O$)—would need to be applied. The calcium would replace the sodium which combines with the sulfate. Leaching with high-quality water is required to move the sodium sulfate downward and out of the root zone. The calcium then helps the soil to reaggregate and improve in physical condition. In some sodic soils, calcium may be present as calcium carbonate and, if so, only sulfur (an acid-forming type) is needed to correct the sodic situation. The general pH management considerations for the soils of the United States are summarized in Figure 9.10.

Biological Characteristics

Life in the soil, or soil biology, was discussed in Chapter 4. One of the principal practices related to the biology of the soil is crop residue management. Crop residue decomposes to replenish soil organic matter, a process that has important and generally beneficial effects on soil biological activity. Other management practices affecting soil biology include the application of animal manure or other organic materials such as biosolids (wastewater treatment residuals) and cannery or processing plant byproducts. Physical or chemical

9.10 The generalized pH management considerations for soils of the United States. Region A soils are generally above pH 7.0-soils may be saline or sodic. In region B the acid-base relationships are commonly favorable, and in region C the bases have been leached so that lime and fertilizer are needed in high amounts.

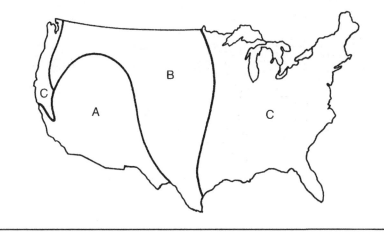

conditions in the soil that adversely affect soil organisms (poor aeration or toxic levels of chemicals) should also be avoided.

Plant residue also serves as a source of energy for organisms that live in the soil. As crop residue decomposes, nutrients that plants require are released. Organic acids are formed that in turn may enhance the availability of certain plant nutrients and also aid in the breakdown of minerals in the soil.

After humus is formed, additional cation exchange capacity is present. Organic matter helps a soil retain more water. Water infiltration rates often increase with additional organic matter as a result of an improved aggregation of soil and enhanced structure. Soil temperature may also tend to be slightly higher because increased organic matter generally causes a soil to darken. Other benefits of increased organic matter include (1) reduced toxicity of certain pesticides, (2) increased buffering in the soil, and (3) a decreased effect of saline–sodic soil conditions.

Crop residue on the surface decreases the detrimental impact of raindrops and thereby decreases erosion caused by runoff water. Residue cover also decreases erosion by wind by slowing the wind near the soil surface. In winter, crop residue can trap and hold snow. As the snow melts, the water moves into the soil where it can be stored and absorbed by the crop later.

In areas of limited rainfall, this extra moisture may be critical for crop production. Surface residue also conserves moisture by keeping soil temperatures lower and decreasing the loss of water by evaporation.

Crop residue can have some potentially detrimental effects. It may harbor diseases that would appear the following year. Weed seeds may also be present and thus may germinate the next year. If the residue is low in nitrogen, which is normally the case with crops like corn and wheat, temporary nitrogen shortages may occur for the next crop to be grown in that soil.

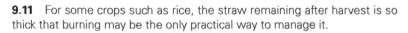

9.11 For some crops such as rice, the straw remaining after harvest is so thick that burning may be the only practical way to manage it.

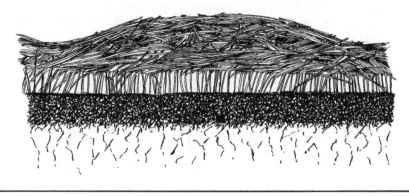

Residue is often burned to avoid one or more of the above-perceived problems. It is undesirable to burn residue because certain nutrients contained in the residue will be lost to the atmosphere and must be replaced through increased fertilization. Burning crop residue is a common practice with rice (Fig. 9.11). In California, some rice straw is baled and injected with ammonia for animal feed.

Crop Production Factors

Crop production is influenced by many factors; some can be controlled by the farmer and some cannot. These factors can be classified as (1) soil, (2) crop, (3) environmental practices, and (4) cultural practices.

The many facets of the soil are discussed throughout this book. Its physical, chemical, and biological characteristics should be maintained under optimum conditions for best crop growth.

The crop to be grown involves many management decisions. A crop needs to be adapted to the soil as well as to climatic conditions. Markets must also be considered. Choice of cropping system (whether to grow one crop continuously or to use a particular cropping sequence) is important. A cropping system may involve continuous corn in the Midwestern United States or a sequence of corn/soybeans every 2 years. In the Southwest, a cropping system may be continuous cotton or a sequence of cotton and some other crop such as grain sorghum or wheat.

Other management decisions on crops include which variety to plant. Some varieties are adapted for high-yielding potentials where fertility and moisture are adequate, whereas others are better adapted to low potential yield situations such as in dryland production areas. The plant population of most crops is important and needs to be maintained at an optimum level.

The environment is generally the most important aspect of crop production and often cannot be controlled or influenced. Length of growing season, altitude, day length, light energy, and rainfall are among conditions over which the farmer has no control. Even though rainfall cannot be changed, certain management practices can be used to conserve water. Contour rows, crop residue cover, furrow diking, and similar practices can be used to increase rainfall retention.

As one moves from the southern to the northern United States, the growing season shortens. Farmers need to plant annual row crops (corn and various bean types) as early as possible to get the highest yield. Therefore, it is very important that the soil temperature is warm enough for seed germination and growth. One way to manage this factor is to move the crop residue from within the row to the inter-row areas of the field. Row cleaners are wheels that travel at an angle to the planting direction and they gently move the residue from the row area to the inter-rows. Having bare soil in the row area allows the sunlight to dry (dry soil warms more quickly than wet soil) and warm the soil which greatly improves early season plant growth.

For some specialty crops, management practices can be used to modify the plant growth environment. For example, additional lighting can be used to control the growing season, day length, and light energy for greenhouse plants, but little can be done for most field-crop production. An exception would be tobacco grown under cheesecloth, for example, where the amount of light reaching the plant is reduced.

Using the right cultural practices at the right time (management) is important. Such decisions start after harvest and could include (1) whether to incorporate residue or leave it on the soil surface; (2) what tillage practices to use; (3) when to plant and how deep to plant; (4) the type and rate of fertilizer and how and when to apply; (5) the rate, time, and method of application of herbicides, insecticides, or fungicides for pest control; (6) when to apply irrigation water if it is available; (7) when to harvest; and (8) what marketing strategy to use. And the alternatives could go on and on. A producer makes many management decisions each year, with almost every decision being a critical one for yield and profit.

Conditions that could limit yield are numerous. A few are cited here as a checklist to determine why yields are less than the potential: (1) biological hazards such as weeds, diseases, or insects; (2) nonbiological hazards such as hail, excessive or insufficient rainfall, early or late frosts, or extreme temperatures; (3) inadequate plant populations; (4) imbalance of plant nutrients or improper nutrient application; (5) poor physical condition of soil; (6) soil pH that is too high or too low; or (7) improper variety (cultivar).

CHAPTER 10

Soil Conservation and the Environment

Conserving the soil has become increasingly important as the need for food and fiber continues to increase. Producers of food and fiber became more aware in the mid-1900s of the need to conserve their most precious resource. Now the public has become more and more aware of the need. And awareness will continue to grow.

Degradation of soil worldwide is apparent. Awareness of this problem is increasing. Those concerned with proper land use and sustained production have called attention to this problem.

Related to both conservation and degradation are environmental concerns. The public has seen the consequences of a lack of conservation resulting in not only the degradation of soil but also the increased detrimental aspects of soil erosion such as pollution, water quality problems, and so on.

Erosion Processes

If rains are gentle on native grasslands and forests, the water soaks into the soil and percolates downward through the soil profile. If there is enough rain, it eventually drains to the water table. When rain falls faster than the soil can absorb it, some water soaks in and the rest runs over the soil surfaces to the lowest part of the field and eventually runs off. The runoff water from vegetated land may look clear, but it is actually carrying some mineral and organic matter derived from exposed soil, earthworm casts, and the digging activities of ants, moles, badgers, and similar organisms.

This process represents normal or geologic erosion. In some places, geologic erosion may lower the ground surface only an inch (2.54 cm) in 5,000 years. This is probably a normal rate for hills and valleys to form. In desert areas where vegetation is sparse and rainfall is infrequent, the rate of erosion is normally slower partly because many desert soil surfaces are often coarse-textured and armored with a layer of pebbles. Geologic erosion includes landslides that occur in mountainous areas as well as the collapse of riverbanks

Soil Science Simplified, Sixth Edition. Neal S. Eash, Thomas J. Sauer, Deb O'Dell, and Evah Odoi.
© 2016 John Wiley & Sons, Inc. Published 2016 by John Wiley & Sons, Inc.

10.1 Loose substratum (A) slowly develops into soil if surface erosion takes place at a slow rate. Where soil is thin over bedrock (B), erosion of the surface leaves a barren landscape.

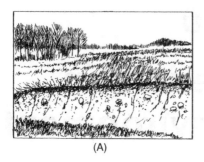
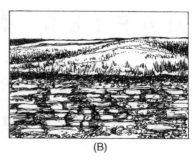

(A)   (B)

into floodwaters during periods of excessively high rainfall. **Geologic erosion** is an ongoing process whereas **accelerated erosion** occurs due to human activities that increase erosion rates. This chapter addresses the impacts of humans on accelerated erosion processes.

In areas where a protective vegetative cover is normal—such as is common in prairies and forests—erosion losses are minimal. Unfortunately, it is usually human activities that cause accelerated erosion due to the removal of vegetation by tillage or overgrazing by livestock. Most agricultural systems tend to accelerate erosion, but it is the obligation of those who manage farms, ranches, and forests to keep erosion to a minimum. High erosion losses degrade streams and rivers, the source of approximately 75% of the drinking water in the United States. This sediment not only impacts aquatic organisms, but also it is a major cost for water treatment plants to remove the sediment prior to water treatment.

European immigrants to the United States cleared the native vegetation to make cultivated fields. Their livestock often overgrazed rangelands, which weakened the plant cover and compacted the soil. As a result, the original condition of the soil deteriorated rapidly and erosion intensified.

The word erosion literally means gnawing away. In regions where the soil parent material is loose, such as where there is thick loess or glacial till, the long-term effects of erosion are serious but not as devastating as where solid bedrock is the parent material (Fig. 10.1). The latter includes most of the earth except for the valleys, which have been filling with sediment over millions of years.

Soil erosion is a serious problem on much of the world's cultivated land (Fig. 10.2). A look at most large rivers shows us they carry a heavy load of sediment that comes from watersheds. A watershed is all the land area that yields water from rain and snow to a particular river. A watershed need not become a "soilshed" too. In arid regions where vegetation is sparse and rains are quite often intense, erosion may be severe without human involvement.

Erosion by Water

Both water and wind erosion consist of three processes: **detachment, transportation, and deposition.** Soil erosion by water occurs when particles are loosened (detached) and carried (transported) by moving water. The soil particles are eventually deposited. As long

10.2 Agricultural systems commonly accelerate erosion.

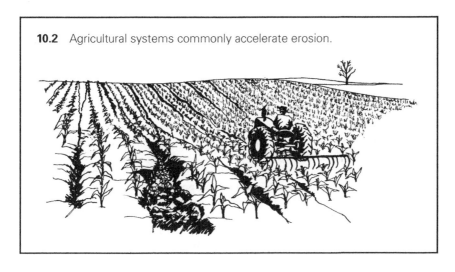

10.3 The impact of raindrops contributes to erosion by breaking up soil aggregates and splashing soil downslope.

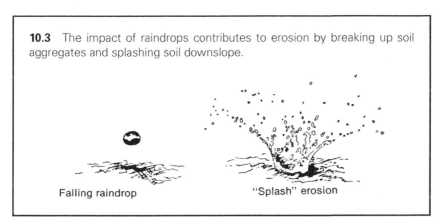

Falling raindrop   "Splash" erosion

as soil particles are clustered into aggregates, they are not easily moved. But if the soil is exposed to the bombarding force of raindrops, the aggregates tend to break apart and the detached particles are subject to movement in the runoff water (Fig. 10.3). Soil aggregates are important to water infiltration and runoff. When the aggregates break apart, the smaller particles plug the larger soil pores important to infiltration. Keeping crop residues on the soil surface protects these aggregates which keep the soil pores open resulting in higher water infiltration rates. As soon as the water infiltration rate is less than the rainfall rate on the soil, runoff occurs.

It has been calculated that the rate of fall of a raindrop is about 20 miles (32 km) per hour. The kinetic energy generated by a 2-in. (5-cm) rain on 1 acre (0.46 ha) is about 6 million foot-pounds (4.4 million joules). This is enough energy to raise a 7-in. (18-cm) layer of soil 3 ft (0.9 m). Clearly, this is not all used to move soil, but it does show that the impact of raindrops releases a large amount of energy that contributes to soil detachment and erosion. On a 10% slope, 60% of the soil splashed by raindrop impact moves downslope and only 40% is thrown upslope. The net movement downhill is called splash erosion.

Types of Water Erosion

The main types of water erosion are gully erosion, rill erosion, sheet erosion, and stream bank erosion.

Sheet erosion is the planing off of a land surface by water action without formation of channels. This generally happens where there is not enough cover of vegetation over the soil to prevent erosion but enough to prevent rilling. Sheet erosion is often unnoticed because only a few millimeters of soil may be lost each year. One way to determine the sheet erosion rate is to find an exposed soil surface and look for a small pebble on the top of the soil. If the pebble is setting on a pedestal, the sheet erosion rate can be determined if one knows when tillage last occurred.

Rill and sheet erosion also often go unnoticed because tillage destroys the evidence until most of the topsoil is gone and the subsoil is exposed at the surface. Although less spectacular than gully erosion, rill and sheet erosion cause the loss of a great deal of more soil.

The most spectacular type of water erosion is gully erosion, which occurs when water concentrates in a channel and deepens it rapidly (Fig. 10.4). Usually a gully is defined as a channel so deep that farm equipment cannot safely cross. This is what happened over a long period to the Grand Canyon of the Colorado River, and it is happening on a smaller scale on thousands of farms throughout the world. A gully generally starts at the outlet of a channel and works its way upstream by waterfall action at its head. The gully extends itself upstream by undercutting the floor of the channel. On a larger scale, Niagara Falls is also working its way upstream, as shown by the occasional fall of masses of limestone bedrock from the rim. By contrast, rill erosion consists of the removal of soil on a side slope by small channels (Fig. 10.5) that are not deep enough to interfere with tillage equipment.

Stream bank erosion is another type of erosion caused by swift moving water that occurs when channels are widened during flood or near-flood conditions. The increased energy of the stream undercuts the stream bank to the point that large blocks of soil actually drop into the channel. The best way to prevent stream bank erosion is to maintain riparian zones with adequate vegetation such as trees and grasses. If stream banks are disturbed through construction, overgrazing, or channelization, stream bank erosion rates skyrocket.

10.4   Gully erosion can be spectacular.

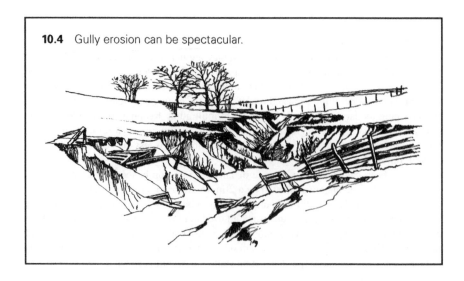

10.5 Rill and sheet erosion can result in great soil loss.

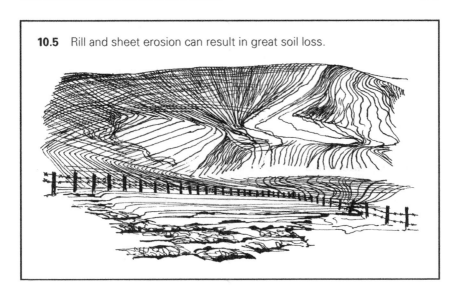

10.6 Two safeguards against soil erosion are vegetative cover and well-aggregated soil.

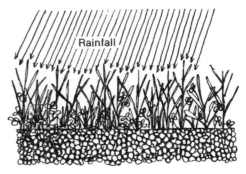

Water Erosion Control

It is an old adage that "an ounce of prevention is worth a pound of cure." This certainly applies to soil erosion control. Because soil particles do not move until they are detached, every effort should be made to prevent this from happening in the first place. Two factors should be kept in mind in this regard. The first is that the strong, water-stable aggregates usually associated with high organic matter content allow water to infiltrate harmlessly into the soil. Second, a cover of vegetation or plant residues functions to dissipate the energy of raindrops so they cannot strike directly on the soil aggregates (Fig. 10.6).

The fundamental principle of preventing the transportation of detached soil particles by runoff water is to reduce the rate of flow down the slopes. Slow-moving water does not have the energy to transport a large load, and the slower speed allows more time for the water to infiltrate.

Contouring and Strip Cropping. One of the common ways of controlling runoff is to subdivide the sloping fields into contour strips (Fig. 10.7). Each strip is approximately perpendicular to the downhill path the water would take. Alternate strips usually contain a row crop such as corn in one strip and a forage or small grain crop in the alternate row. The ridges in the strip crop reduce the water flow and minimize erosion there, and little erosion can occur in the dense plant cover in the alternate strips. No-till planting, in which the previous year's crop residue is left on the surface, is another practice that reduces erosion even when it is not done on the contour. It permits use of agricultural equipment that is too large for contour strip-cropping operations.

Grass Waterways. Where water concentrates in channels and flows in an uncontrolled manner, the channels may become gullies. Grass waterways (Fig. 10.8) can reduce the risk of gully formation because the dense fibrous root system of grasses holds

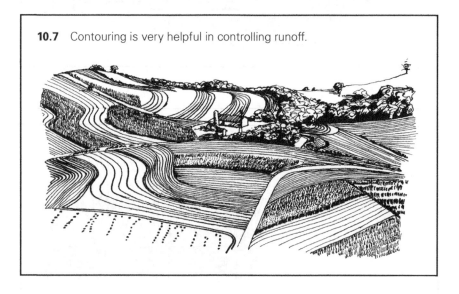

10.7   Contouring is very helpful in controlling runoff.

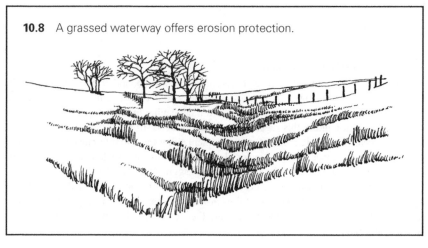

10.8   A grassed waterway offers erosion protection.

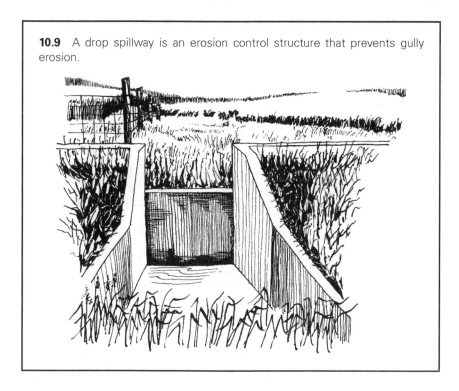

10.9 A drop spillway is an erosion control structure that prevents gully erosion.

the soil securely against the force of the flowing water. Grass waterways may be planted in natural channels that lead downslope, in diversion ditches on the sides of slopes, and at the foot of adjoining slopes.

Conservation Structures. If the natural slope of the channel is too steep, it can be flattened out between steel or concrete structures called drop spillways (Fig. 10.9), which act as steps in the channel. At each step, the water drops over a headwall and falls on a concrete apron from which it flows onward through a grassed waterway.

In commercial agriculture today, conservation practices have advanced to the point that gullies are seldom a major problem. The size of farm machinery is such that if gullies start to form, they are easily removed. If gullies continue to form, however, farmers often are forced to divide their fields inconveniently because large gullies cannot be crossed with machinery. Fences are undercut, and roads have to be rerouted around advancing gullies. In some instances buildings have toppled into them.

To stop the advance of a gully, the soil must be stabilized at its head where the waterfall action is occurring. Sometimes the head of the gully is filled in with concrete or rubble (riprap), and the sides and bottom are planted with trees, grass, and shrubs. These practices may be combined with the use of a chute through which runoff water passes to prevent any further undercutting at the gully head. Various other techniques may be employed, depending on the specific problems at hand.

Contour Ridges and Terraces. A principal point to keep in mind is that in many cases it is desirable to keep as much of the water as possible where it falls or is applied by irrigation so that it can move into the soil and be used by plants. This is very important in arid and semiarid areas where moisture is often the limiting factor. Contouring and strip

10.10   Furrow dikes trap most of the water that falls as rain or by sprinkler irrigation so it can be used by the crop.

cropping are techniques in this category. Another approach is furrow diking, which is an old practice that is now being used much more commonly in dryland areas (Fig. 10.10). A dike is placed in each furrow at intervals of 6–12 ft (1.8–3.6 m), with a resulting depression holding water between the dikes. This greatly decreases runoff and improves plant growth. In the furrows the tractor follows during harvest; the dikes must be removed by sweeps mounted in front of the tractor.

Another mechanical erosion control practice is terracing. There may be no older erosion control method than the construction of terraces some of which are more than two thousand years old. In some parts of the world, entire mountainsides have been modified into a series of steps by bench terraces, each one supported by a retaining wall made of stones (Fig. 10.11). Tremendous amounts of labor were required to construct them, and they still require labor to maintain, which is very cheap in some areas. Terraces on a much more modest scale have been used compatibly with modern agriculture. In these cases, ridges are built several yards (meters) apart across the slope of the hills to trap water falling on the soil between the terraces. Presently, the trapped water is usually allowed to flow down the hill through buried tiles so that erosion does not occur (Fig. 10.12). In drier areas, the water may simply be held above the terrace until it soaks into the ground.

Erosion by Wind

Whereas the loss of soil by water erosion is more widespread than by wind, it was the devastating wind erosion of the drought years of the mid-1930s in the United States that focused attention on all types of soil erosion and brought about legislation dedicated to soil conservation (Fig. 10.13). In the late part of the nineteenth century and in the early twentieth century, much of the prairie in the Midwest was plowed with huge steam tractors and planted to wheat. In addition, much of the unplowed land was overgrazed so that the soil was also left with an inadequate vegetative cover to afford protection. When the drought and

10.11   Agricultural terraces for rice production in the highlands of Vietnam. These terraces are irrigated and have been productive for more than 1,000 years.

10.12   Parallel terraces may be drained by buried tiles.

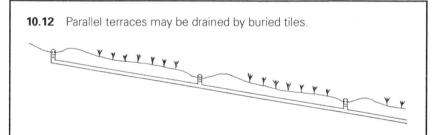

winds came in the 1930s, soil from the Great Plains was blown aloft in ominous dark clouds that sometimes reached the Atlantic Coast and even out to sea. Fences were covered, road ditches were filled, and farmyards were smothered with eroded soil, and there was much despair in the hearts of farmers and city folks alike. Even today, wind erosion continues to move vast quantities of the most fertile topsoil, thereby reducing the productive capacity of cropland.

Types of Wind Erosion

Wind erosion occurs when the wind is sufficiently (a) strong to move soil particles along the soil surface; (b) turbulent to keep particles suspended; and (c) gusty to keep soil moving. There are three processes that occur during wind erosion events: saltation, surface creep, and suspension (Fig. 10.14).

10.13 Although less common than water erosion, wind erosion can be devastating.

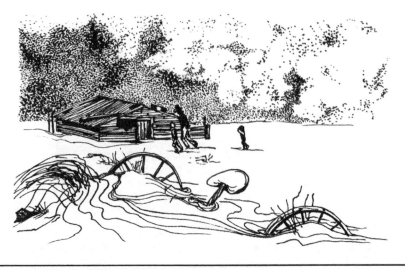

10.14 Wind erosion transports soil particles by creep, saltation, and suspension.

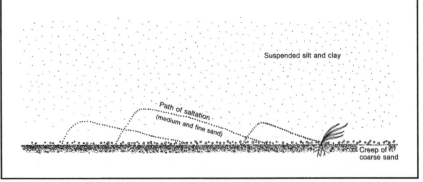

Suspended silt and clay

Path of saltation
(medium and fine sand)

Creep of
coarse sand

Saltation is the bouncing of medium and fine sand particles along the surface after they initially start to roll. The bouncing sand particles hit other soil particles which then project other particles (sand, silt, or clay particles or small soil aggregates) into the air. **Surface creep** is the rolling of coarse sand particles on the soil surface. Damage occurs to growing plants when surface creep and saltation occur. Sand particles constantly bombard the growing plants and eventually kill them if the wind is severe. Row crops less than about 4 in. (10 cm) tall can be cut off by the blowing soil. To protect the crop, farmers will often till strips on the ridges of the field to bring moist soil to the surface in the hope that the high winds will be temporary.

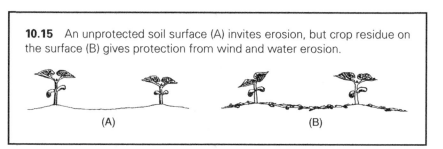

10.15   An unprotected soil surface (A) invites erosion, but crop residue on the surface (B) gives protection from wind and water erosion.

(A)   (B)

The abrasive action of the sand particles loosens finer soil particles, silt and clay, and brings about the third type of wind erosion, **suspension.** These fine soil particles, silt and clay, are often suspended in the air, resulting in a dust storm, and may be carried hundreds of miles. Although this may not be too damaging to growing crops, the loss of valuable topsoil with its organic matter and nutrients is costly. Damage may also occur when the soil particles are deposited. Deposition may be at the edge of a field where sand is deposited or many miles away from the origin of the particles. Recent studies have shown that dust from deserts in Africa blows across the south Atlantic and supplies nutrients for plant growth on otherwise infertile soil in the Amazon basin of South America.

Control of Wind Erosion

Vegetative Covers.   Control of erosion by wind is achieved by slowing down the wind at ground level, much the same as controlling erosion by water is accomplished by slowing the runoff. Vegetative cover is the most effective means of control by simply keeping the soil covered so that it cannot be blown away. The vegetation may be alive in the form of a growing crop cover, or it may be dead plant residue sheltering the soil (Fig. 10.15). Often crop residue of this nature is partially incorporated but a portion is left on the surface to prevent blowing of the soil. The common tillage method to achieve this stubble mulching is shown in Figure 10.16.

Roughened Soil Surface.   Leaving the soil in a roughened condition also helps control wind erosion. Any obstacle in the way of a moving soil particle deters it and, in many cases, stops it. The obstacle might be cloddy soil, pitted areas in the soil surface, or ridged rows that have been prepared. Farmers in the Great Plains of the United States use a sand fighter, which is an implement taken over a field after a rain when the soil surface is smooth to create thousands of small pits and mounds to impede the movement of soil. This is usually done in the spring when wind speeds are highest and crops have just been or are soon to be planted. Using a sand fighter may kill a few plants if the crop is up, but damage is minimal.

Strip Cropping.   Strip cropping is another means of minimizing soil movement by wind. The pattern would be similar to that used for erosion control by water, such as preparing a seedbed in the standard method for 8–12 rows but leaving 1–2 rows of the old crop standing. Strip cropping is often used on vegetables grown following a volunteer wheat crop.

Windbreaks.   Windbreaks, or shelterbelts (Fig. 10.17), are used in many areas where erosion by wind is a problem. A planting of trees, shrubs, or grass strips helps to slow the wind and decrease soil movement. To be effective, the distance between windbreaks

10.16 Stubble-mulching loosens the soil but leaves most of the plant residue on the surface.

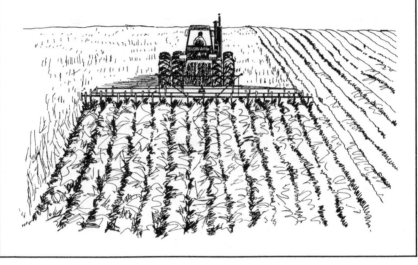

10.17 Shelterbelts and rough soil surfaces can reduce wind erosion.

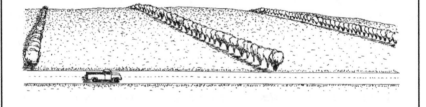

should not be too great. The distance for best control varies from one area to another. Windbreaks are used less today due to the advent of larger machinery for tillage and harvest operations. Farming around windbreaks takes more time and the trees in the windbreak consume soil moisture needed by the crops. Contact your local Natural Resources Conservation Service personnel for specifics.

    Soil Stabilizers.   Soil stabilizers might also reduce erosion by wind or water. These include organic matter that acts as a binding agent, chemicals that cause soil particles to aggregate, and water. Water stabilizes the soil for the time when it is moist, but when the surface dries out, it is again susceptible to wind erosion.

Erosion by Mass Wasting

    Mass wasting is another type of erosion that takes place in several ways. Masses of soil sometimes move under the force of gravity, as by cave-ins along riverbanks (Fig. 10.18) and slump along gully sides and road banks. Soil creep is a common type

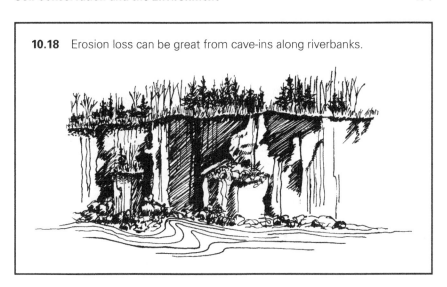

10.18 Erosion loss can be great from cave-ins along riverbanks.

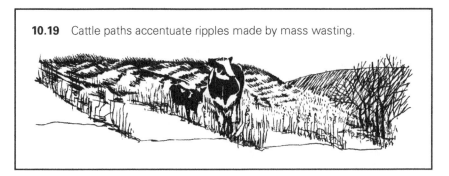

10.19 Cattle paths accentuate ripples made by mass wasting.

of mass wasting in regions where the soil freezes and thaws repeatedly each autumn and early spring (Fig. 10.19). Frost causes the surface soil to expand perpendicular to the slope; when the soil thaws, it drops vertically. This causes a ripple effect on the hillsides. Cattle often make paths that accentuate these ripples. Mass wasting not only may remove portions of cultivated fields and pastures but also sections of roads and even homes (as in the mudslides in California).

Diversion of water away from susceptible areas is helpful. Protection of riverbanks with riprap, timberwork, and vegetation reduces the incidence of mass wasting. In some cases, channelization is needed but, in general, is only a temporary solution. Similar protection is also being sought for shorelines and beaches; huge sand-filled plastic tubes are laid at the foot of an endangered bluff at a lake or seashore to block the force of waves. The city of Miami, Florida, has spent huge sums of money trying to decrease beach erosion along much of its waterfront.

Whenever soil is moved by wind, water, or gravity, it is likely to cause immediate problems in addition to those of a more long-range nature. The occurrence of massive mudslides in such places as California usually attracts widespread media attention. This accelerated erosion is often attributable to the loss of vegetation due to fire or logging. Houses may be filled with the flowing ooze or even swept down the hillsides.

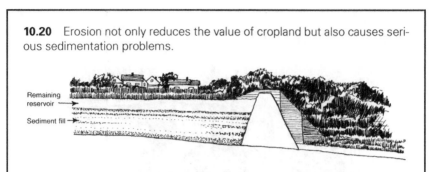

10.20 Erosion not only reduces the value of cropland but also causes serious sedimentation problems.

Remaining reservoir ⟶
Sediment fill ⟶

Sediment as a Pollutant

Eroded soil from uplands is carried by rushing water until the water slows down and drops its load of sediment in river channels, harbors, and reservoirs (Fig. 10.20). Dredging these sediments to remove the sediment is costly. Depleted channels increase the threat of flooding, which causes untold hardships to homeowners and businesses. Dams built to control floodwaters and provide other services such as recreation and electric power lose their value as the capacity of their reservoirs decreases. In extreme cases, expensive new dams have become a burden to the taxpayer in as little as 25 years.

Sand can be a sediment since it fills ditches, covers roads and fences, and causes similar problems. When it is blown along by the wind, it is a serious problem, whether in central Wisconsin or in New Mexico. Valuable topsoil is lost and productivity declines, and small plants are cut off or abraded so that they fall prey to disease. The abrasive action of sand can remove paint from buildings and automobiles, and result in textural changes in soil surface horizons. This loss of soil is also a loss of valuable plant nutrients and can carry pesticides into receiving streams and lakes.

It should be apparent that soil conservation pays. It pays where the soil is retained because the land can be more productive, and it also cuts costs downstream where the capacity of reservoirs is protected and water quality for domestic, irrigation, and recreational uses is preserved.

One of the few benefits of sedimentation is that it results in the formation of alluvial soils in flood plains and at the mouth of rivers such as the delta of the Mississippi River.

Extent of the Problem

A natural resource inventory reported by the Natural Resources Conservation Service (NRCS), in 1992, revealed that the national average soil loss per year on all cropland due to water erosion was 3.1 tons per acre (T/A) equal to 6.9 megagram per hectare (Mg/ha) per year. The figure varies for different parts of the country, ranging from 6.2 tons per acre (13.9 Mg/ha) per year in Alabama to 0.1 (0.2 Mg/ha) in Colorado.

Many highly eroded areas experience losses in excess of 20 tons per acre (45 Mg/ha) per year. Drier regions with low amounts of water erosion may experience losses from wind erosion that match or exceed these values. Nationwide, wind erosion losses average 2.5 tons per acre (5.6 Mg/ha) per year. Comparative values for years prior to 1992 were 3.2 T/A

(7.2 Mg/ha) in 1987 and 3.3 T/A (7.4 Mg/ha) in 1982. NRCS reported the 1992 wind erosion figures for the individual states. The highest wind erosion occurred in New Mexico, which lost 13.5 T/A (30.2 Mg/ha) per year, while several states in the higher rainfall areas showed no wind erosion losses. Some of these states are Alabama, Arkansas, Connecticut, Georgia, Hawaii, and Kentucky.

To place these figures in perspective, it is worthwhile noting that 1 in. (2.54 cm) of soil weighs about 150 tons (136 Mg). By using some of the soil-loss figures cited above, it is easy to calculate how long it takes to lose an inch of topsoil at those rates. For example, at 5 tons per acre (11.2 Mg/ha) per year, 1 in. (2.54 cm) of soil is lost every 30 years.

It is important to remember that eroded soil is not necessarily removed from a farm and transported to a lake or to the sea. It is more likely to be shifted from the high ground to the low ground. The high ground becomes much less productive, and crops on the low ground may be damaged by sediment and eventually those soils may become more productive.

Despite the great work of researchers and government agencies, the NRCS in particular, only about one-half of the farmers in the United States have requested conservation plans for their land, and only about one half of those have carried them out. Some of the best-managed farms may have changed ownership and resulted in the soil conservation practices being abandoned by the new owners.

An excellent US Department of Agriculture (USDA) publication by W. C. Lowdermilk in 1953, "Conquest of the Land through 7,000 Years," cites many examples of Old World states whose foundations were based upon agriculture and whose demise came when their abused land was no longer productive. In some cases, population increases placed too high a demand on the fragile soil, and in others, political instability led to neglect of both the soil and the irrigation systems.

It may appear to some farmers today that soil conservation really is not very important. For instance, a yield of 200 bushels of corn per acre (12.5 Mg/ha) can be achieved on the black prairie soils of Illinois under an annual rainfall of 40 in. (1,000 mm). Wherever soil deteriorates by erosion (and that is just about everywhere in cultivated fields) and by compaction, which reduces soil aeration, farmers can still obtain high yields by adding more fertilizer, by planting more vigorous hybrid corn varieties, and even by irrigating in seasons of low rainfall. Thus good crop yields are achieved decade after decade, while the soil is slowly but surely wasting away unnoticed. News reports of crop surpluses in the United States and many other countries lull the public and politicians into complacency on the state of the soil resource.

Our U.S. population continues to grow. With this growth there have been astronomical rates of urbanization that destroys many acres of prime farmland each day. Hopefully, it will become a national priority to keep our prime farmland growing needed food crops and not growing houses and septic tanks. If current population growth continues, this land will be needed to feed future generations.

CHAPTER **11**

Conservation Agriculture

Conservation Agriculture (CA) is an agricultural management approach that evolved to sustainably address issues of soil erosion as described in Chapter 10 and soil management as described in Chapter 9 for crop production. The Food and Agriculture Organization (FAO) of the United Nations (UN) defines CA as an approach to farming (crop production) that sustains and improves production for both food security and profits while also preserving and enhancing environmental resources (United Nations, 2015). CA is based on the following three principles:

1. Minimum soil disturbance,
2. Permanent soil cover, with crop residues, mulch, and/or cover crops,
3. Crop rotation/diversity,

These principles are widely applicable to many different crops, climates, and soils; for example, they have been successfully applied in a range of climates from the Arctic Circle to the tropics.

CA Principles

Minimum Soil Disturbance
Planting. Cultivation of crops requires planting of seeds. There are many approaches to planting seeds, from direct seeding where seed is dropped into a slot that is then pressed closed with minimal soil disturbance to intensive tillage that prepares a "clean" seedbed with primary tillage or inversion of the top layer of soil with a moldboard plow followed by secondary chisels, disks, and harrow tillage resulting in a cycle of erosion that has severely degraded soils. CA advocates minimum soil disturbance, where crop seeds are inserted directly into the soil through the residue of the last crop, omitting the

Soil Science Simplified, Sixth Edition. Neal S. Eash, Thomas J. Sauer, Deb O'Dell, and Evah Odoi.
© 2016 John Wiley & Sons, Inc. Published 2016 by John Wiley & Sons, Inc.

steps of seedbed preparation with intensive tillage. Direct seeding/sowing, direct drilling, no-till farming, zero tillage, and no-tillage are all terms used to describe this approach. Large seeds such as maize and beans can be placed at a desired depth with mechanized equipment called no-till planters or drills that cut through the residue, open a slot and place the seed into that slot, ideally closing the slot so that the seed is in good contact with the soil and the soil is covered by the residue. Very small seeds may be broadcast and will germinate without incorporation just as occurs in native ecosystems.

If tractor-drawn planters are not available, no-till or minimum tillage can also be carried out manually, typical of small scale farming, using several methods including planting basins, planting spots, rippers, planting sticks or jab-planters, and animal-drawn planters. A planting basin is a hole dug 6 in. (15 cm) wide by 12 in. (30 cm) long by 6–8 in. (15–20 cm) deep and spaced in rows about 30 in. (75 cm) apart, using a hoe as the digging implement. Basins are prepared during the dry season and after being dug, inputs such as manure, fertilizer, and lime can be placed in the basin, which is then partially filled with soil to the planting depth. At seeding time after the first rain, multiple seeds are placed in the basin and then covered with soil and/or mulch. Shortly after germination the plants are often thinned to 2–3 plants per basin. After basins are dug the first year, they can be reused in subsequent years thus saving labor. Basins direct rainfall to the seed area and provide a framework for more precise placement of fertilizer.

Planting spots are more shallow planting holes dug to the seed planting depth with a hand hoe, sowed with seed, and then covered with soil, and can be planted with less time than basins in the dry season or just after the rains begin. About 2 weeks after the crop emerges, a stick can be used to make a hole about 2.5 in. (10 cm) away from the seedling to place fertilizer for plant growth.

A wooden stick with a pointed end, called a planting or dibble stick can be used to create a hole to a certain depth. The seed is dropped in the hole, which is then stepped on to close the hole and ensures seed contact with soil. Jab planters are light-weight, hand-carried, and -operated devices that have a seed hopper connected to a shaft that delivers 1–3 seeds at a time to a steel furrow opener at the end of the shaft. The action of pulling the two handles apart and simultaneously jabbing the tip of the shaft into the soil, sets the seed into the tip. A second action of pulling the handles together delivers the seeds and/or fertilizer into the hole created by the tip. Jab-planters can come with one or two hoppers for seed and fertilizer. Jab planters can also be used to fill gaps that are noticed after germination.

Rippers are chisel-pointed tools that open a narrow slot (furrow) about 2–4 in. (5–10 cm) deep for sowing seeds by hand, a machete or planting stick. A mechanical planter attached to the ripper itself can also be used to insert the seeds into the slot. Ripped lines are usually spaced about 30 in. (75 cm) apart.

Several varieties of animal-drawn, hand-operated, automatic, and semi-automatic planter devices are available for purchase and use that can reduce the time for seeding.

Weed Control. Weeds compete with crops for moisture, nutrients, space, and sunlight. Weeds can also harbor disease and pests that can attack crops. Traditional tillage agriculture used the plow as the major weed control technique incorporating weeds and residue into the soil, providing a clean seed bed in which to plant and also providing nutrients for the subsequent crop from the incorporated plant residues. In CA, weeds are killed and left in place using herbicides, by hand weeding or using equipment to cut or crush weeds, avoiding soil disturbance as much as possible and leaving plant residue on top of the soil.

Herbicides do not disturb the soil and can be quick and easy to apply. Herbicides require special application equipment such as sprayers or wipers and they require knowledge or training to determine which herbicides are appropriate and how to prepare, handle, and apply them correctly and safely. Herbicides may not be accessible or affordable by smallholder farmers.

Weeding with hand tools or hoes can be used to manually dislodge weeds. Though hand weeding is harder work, it generally disturbs the soil less than using a hoe. Knife rollers kill weeds and cover crops by bending and crushing them, and can be used before seeding the crop.

In CA, weeds are also controlled by planting crops closer together to shade out weeds. Ideal crop spacing depends on soil moisture, temperature, and fertility, and has the objective of avoiding competition between the crop plants for water and nutrients. Weeds are also controlled using cover crops, mulch, and crop rotations as described in the following sections.

Permanent Soil Cover with Crop Residues, Mulch, and/or Cover Crops

Making sure the soil is continuously covered with the main crop, mulch, crop residues, and/or cover crops protects the soil from the eroding force of raindrops and excessive heating by the sun. Preventing the eroding force of raindrops on exposed soil reduces soil crusting and surface sealing enabling greater rainfall infiltration and reduced surface runoff and flooding. Constant soil cover also reduces soil moisture evaporation losses thereby increasing soil moisture content available to growing plants during dry periods.

There are two main types of soil cover, including: (1) living plant material such as crops and cover crops; and (2) mulch, compost, and/or dead plant material, which includes crop residues and cuttings or leaves from grass, shrubs, and trees. A combination of mulch and cover crops may be used to keep the soil covered.

By reducing the force, speed, and splash effects of raindrops, residues, mulch, and cover crops allow higher infiltration of water into the soil and reduce runoff, which also decreases soil erosion. The residues also form a rough physical barrier that reduces the speed of water and wind over the surface. Reducing wind speed decreases evaporation of soil moisture.

Soil cover provides the following benefits:

- Food and habitat for soil micro- and macroorganisms (Chapter 4), which perform important roles decomposing organic matter, nutrient cycling, soil mixing, and the development of soil pores and structure.
- Insulation from maximum heating and cooling providing a more temperate microclimate for optimal growth of soil organisms and plant roots.
- Weed suppression by competing with weeds, restricting sunlight, and reducing weed seed germination.
- Improved water infiltration and retention of soil moisture, making more water available to crops over a longer period and increasing availability of plant nutrients. By increasing water infiltration, water runoff, and erosion are reduced.

One of the biggest general benefits of CA may be reduced soil erosion. Erosion of soil into waterways leads to filling water reservoirs, lakes, and streams with sediment, reducing water storage capacity and the useful life of reservoirs and dams. Sediment in surface water increases wear and tear in hydroelectric installations and pumping devices, which result in higher maintenance costs and necessitates earlier replacement.

Because more water infiltrates into the soil with CA rather than running off the soil surface, streams are fed more by subsurface flow through soil than by surface runoff. Thus, surface waters are cleaner and more closely resemble groundwater in CA than in areas where intensive tillage and accompanying erosion is more prevalent. Greater infiltration will reduce flooding, by causing more water storage in soil and slower release to streams. Infiltration also recharges groundwater, which can increase water supplies of wells and springs. Sediments in surface waters have to be removed from drinking water supplies, so a reduction in eroded sediment in streams can lead to lower costs for water treatment.

A cover crop can be planted during the cropping season in between the crop rows, known as "intercropping." For example, a low growing plant such as beans can make a good intercrop with a tall plant such as corn. Cover crops may be planted following the growing season to cover the whole field, though in arid or temperate climates with cold winters, it may be hard to grow or maintain a cover crop during the non-growing/dry season. A cover crop can also be planted in the main crop residue towards the end of the growing season, which gives the cover crop a head start to grow at the beginning of a cold or dry non-growing season.

Cover crops have multiple uses, including as edible seeds or vegetables, as animal fodder, firewood or fencing material, medicine, or to increase soil fertility as with nitrogen-fixing legumes. A cover crop that is used to add nutrients such as nitrogen to the soil is known as a green manure cover crop (GMCC). A cover crop is selected based on how it addresses the needs of the farmer or cropping system.

Some cover crops may be harvested for food or as cash crops such as winter wheat (Triticum aestivum). Cover crops may also be selected based on their ability to produce high amounts of residue for fodder or for strong root development that can pierce hard pans and reduce soil compaction, such as radish (*Raphanus* spp.) or pigeonpea (*Cajanus cajan*). Some cover crops, such as marigolds (Asteraceae), have insect pest repelling capability, or weed suppression capability with allelopathic compounds, such as ryegrass (*Lolium perenne* L.), which can inhibit the germination and growth of some weed species.

Crop Rotation

Crop rotation is a key principle of CA because it helps to control weeds, diseases, and pests and it helps to improve soil fertility and structure. Crop rotation involves the diversification of crop species and the sequence they are grown, or the variety of crops grown together (in association) for perennial species. Rotating crops increases the biodiversity of the soil environment and reduces the carry over and growth of crop-specific pests and diseases that result from planting the same crop in succession (monocropping). Crop rotation increases biological activity in soils by providing different food sources and diverse rooting structures. Some crops have strong deep roots that allow nutrients and moisture to be extracted from deeper layers in the soil, while other plants have shallow fine roots that create structure and channels at higher layers. These different types of roots enable nutrients to be extracted and recycled over a larger area. Different plants extract and store different nutrients, so varying the crops planted improves the distribution and balance of the major plant nutrients (nitrogen, phosphorous, and potassium) and minor plant nutrients available in the soil, contributing to diversity of the biota above and below the surface.

Crop rotations are planned according to objectives of food, fodder, and residue production; pest and weed control; and nutrient uptake and production. Some crop rotations may have a cycle of seven or more different crop species in a sequence. Species diversity can be achieved by planting a mixture of cover crop species in one season. Crop selection is also based on soil and climate conditions. Crop rotations, multiple crops grown in a year, and/or intercropping can reduce risk, since a single crop can fail due to a drought or attack by pests.

Crop rotation helps to replace plowing of the soil by aerating the soil with different types of roots, adding the organic matter from roots at various depths in the soil, recycling nutrients, and controlling weeds, pests, and diseases that can live in the residue and soil.

Synergies between the Principles

Though these are well known principles, in CA they are combined to work in concert, for example, maintaining soil cover and reducing soil disturbance by not tilling reduces the erosion that results in loss of soil, reduced soil fertility, and soil compaction. All three principles work together to increase biodiversity and soil organic matter, which increases soil fertility. These principles capitalize on natural biological, chemical, and physical processes above and below ground, especially the physical, chemical, and biological properties of organic matter that holds water and nutrients like a sponge, thereby supporting both plant and animal life and natural ecosystem services.

The constant addition of crop residues from not tilling and keeping continuous soil cover with residues, mulch, and cover crops lead to an increase in the organic matter content of the soil. In the beginning this is limited to the top layer of the soil, but with time this will extend to deeper soil layers. Organic matter plays an important role in the soil; fertilizer use efficiency, water holding capacity, soil aggregation, rooting environment, and nutrient retention all depend on organic matter.

In CA soil plants' roots and macrofauna such as earth worms perform "biological tillage" also called soil bioturbation, in place of the mechanical tillage of plows. Charles Darwin estimated in his book entitled: *The Formation of Vegetable Mould Through the Action of Worms With Observations of Their Habits* (1881) that earthworms could turn over 16 tons of soil per acre (40 tons per ha) per year, and Darwin points out with great foresight that:

The plough is one of the most ancient and valuable of man's inventions; but long before he existed the land was in fact regularly ploughed, and still continues to be thus ploughed by earthworms. It may be doubted whether there are many other animals which have played so important a part in the history of the world, as have these lowly organized creatures.

Though all three principles are generally required to realize optimal results, often CA is adopted in a step-wise fashion with one or two principles. Though it may appear easier, for example, to implement no-till to start, there can be a large cost such as the proliferation of weeds and pests when only no-till is applied.

CA Adoption

History of CA

The events leading up to the development of CA began with the Dust Bowl in the Great Plains of the USA. Following the U.S. Federal Homestead Acts in the late 1800s, which settled much of the Midwestern plain states, and the increase in wheat (*Triticum aestivum* L.) prices during World War I, farmers plowed millions of acres with new gasoline tractors in the southern great plains states of the US through the 1920s, referred to by many as the "great plow-up." When an 8-year drought began in 1931, agricultural land dried up and 100 million acres of plowed, unprotected cropland lost most or all of its topsoil from wind erosion. The U.S. Soil Conservation Service, established in 1935, addressed soil erosion with concepts to protect the soil including windbreaks, contour strips, terraces, grassed waterways, and contour plowing, the first change in conventional tillage.

One of the first to perceive and publicly question the damage caused by tillage using the moldboard plow was Edward Faulkner in his book *Plowman's Folly* (1943), paving the way for others to seriously consider and explore no-till. Simultaneously Masanobu Fukuoka began experimenting in Japan with no-till concepts; however, his work *One Straw Revolution* (1978) did not get global attention immediately.

After the invention of the herbicides 2,4-D, atrazine, and paraquat in the 1940s and 1950s, no-tillage research gained momentum in the 1960s in the USA and the UK, and the first mechanized demonstration farm trials showed effective use of no-till in 1961. With the development of equipment, several US farms and universities demonstrated successful applications of no-till with a variety of crops, and US universities set up Extension Programs to promote no-till. Equipment manufacturers began selling no-till planters in the mid-1970s, which made no-till practical to adopt by large and medium scale farmers.

No-till also gained a foothold in Brazil, due to a government policy encouraging a shift from livestock farming to cropping systems in the high rainfall hilly areas in southern Brazil, to take advantage of rapid growth in global demand for soybeans in the 1960s. Farmer use of the plow had produced severe soil erosion, which dramatically reduced yields. A Brazilian farmer contacted the University of Kentucky in the early 1970s, which provided access to early no-till equipment, launching a collaboration in Brazil between farmers, researchers, and equipment manufacturers. During the 1980s farmers and researchers supported by the Brazilian government and industry adapted equipment for clay soils; and from their experiences added the practices of rotation and cover crops to no-till to form the basis for the three CA principles. This was first adopted by larger Brazilian farmers in the 1980s and then smaller farmers in the 1990s.

From the 1970s to the 1990s, farmers, researchers, and equipment manufacturers in Brazil and the USA developed and advanced no-till farm equipment and management practices to improve performance of field operations and crops, reaching a base of adoption that enabled demand and corresponding supply by industry of no-tillage farm equipment.

With a lot of the earliest research in to-till taking place in the USA, the largest no-till area in the world as estimated in 2009 was in the USA with 88 million acres (35.5 million ha), though the percentage of cropland under no-till in the USA is 35%, and only 10% of US cropland is under continuous no-till, with the other 25% using some form of tillage, like strip tillage. Table 11.1 shows the countries with the greatest percentage of cropland under CA as reported to the FAO between 2008 and 2014 (FAOSTAT, 2014). As of 2007, about 76% of global land under no-till was in the Americas, 12% in Australia and New Zealand, 5% was in Asia, 4% in Russia and Kazakhstan, with only 1% in Europe, and 1% in Africa.

Table 11.1 Area of land under no-till in the top countries as reported to FAO from 2009 to 2014

Country	Million acres	Million hectares	Area of CA as % of total arable land
USA.	88	36	22
Argentina	67	27	71
Brazil	79	32	44
Australia	44	18	36
Canada	45	18	39
Paraguay	7	3	62
Uruguay	3	1	36

Though initial research also took place in the UK and other countries such as Nigeria and Kenya, adoption has been slow in Europe and Africa due in large part to traditional farmer practices.

The UN FAO began formally promoting the adoption of CA around the world since 2002 and provides an extensive source of information, educational materials and adoption status on its website (at: www.fao.org/ag/ca).

Challenges to Adoption of CA

Making a major change to a farming operation can have high startup costs including planting equipment and the time required to learn and effectively apply a new system. Growing crops and managing a farm is complex and despite the effectiveness and relative simplicity of CA, applying the three principles depends on the conditions and must be tailored to the specific climate, crop, and soil. In addition to the crop and environmental setting, agriculture depends heavily on management decisions such as the planting populations and the timing of planting, fertilizer applications, and weed control. Effective management decisions are crucial for achieving optimal yields with CA.

Because tillage mineralizes incorporated organic matter through microbial decomposition, it provides a flush of nutrients similar to a fertilizer application. This has a short-term benefit for the immediate crop, but over the long-term tillage reduces this nutrient stock through the loss of soil and organic matter content from erosion and decomposition, while CA builds the organic matter stocks and reduces erosion. Because of this short-term effect, tillage has been associated with increased fertility and has had the appearance of being a valuable tool, until most of the organic matter is lost and the soil is degraded. Without adding nutrients back into the soil, such as with compost, mulch, manure, and/or fertilizer, crops will not have sufficient nutrients. This soil degradation resulting from tillage may be one reason for the historical use of fallow periods and shifting agriculture to enable natural ecosystems to restore nutrients through re-vegetation.

Adding nutrients such as fertilizer, can often make up for the loss of organic matter and nutrients in degraded soils, however, this is often costly. Fertilizer also does not make up for the many positive qualities of organic matter, such as water holding capacity. Even with fertilizer additions, over time farmers find that their crop yields decline with degrading soils.

Farmers used plowing and burning of plant residues to increase fertility and to control weeds, disease and pests. So, one of the greatest barriers to adoption of CA is changing farmers' intuitive understanding about tillage. Disease and pests can increase with the change from tillage agriculture to no-till. Managing pests and weeds without plowing or burning is not simple or obvious. Integrated pest management is critical for success in CA, and it requires knowledge and training. Crop rotation and diversity halts the growth of specific pests and plant diversity takes advantage of the chemical and physical interactions of different species. While synthetic pesticides, especially herbicides are indispensable especially in the beginning, after beneficial organisms become established and organic matter is increased, it is possible to reduce the use of chemical pesticides, herbicides, and fertilizers.

Because the change from conventional to CA requires significant startup costs, is a change from deeply ingrained traditions, has a steep learning curve, and most often produces greater weeds and lower yields in the first few years, there is a significant barrier to adoption. Weeds are noted as one of the biggest problems and can take several years to get under control. Also during the transition to CA some pests and disease can create problems until a more diverse biological community takes hold. The main benefits of CA may take from 3 to 7 years to be realized (Pope, 1989).

Erosion and severely degraded soils in South America and in the southern USA helped to spur adoption of no-till in those areas, because decreasing yields on eroded and degraded soils represented a greater cost. By offering subsidies, the government of Brazil was able to promote adoption by farmers and the development of a no-till equipment industry. Without educational programs and government support, it may be difficult for the average farmer to overcome the barriers to adoption.

Also, there is no exact recipe for applying CA to a specific location (soil and climate). Achieving a new farm ecosystem balance requires farmer observation and testing, that is, adaptive research on the farmer scale and the sharing of knowledge gained between local farmers. Realizing many of the benefits of CA, can take time, even decades. Apprehension about converting from tillage agriculture to CA can be overcome with the formation of farmer communities that provide a forum for the exchange of ideas and knowledge among farmers practicing CA, building a foundation of local knowledge.

Though continuous no-till provides minimum soil disturbance, some farmers that have adopted CA will occasionally till for various reasons such as incorporating lime in dry climates, or combating herbicide resistant weeds, increasing nutrient mineralization, or to control certain pests.

Trends in No-Till Adoption in Mechanized Agriculture

In 2012 the USDA estimated no-till practices at 92 million acres in the US. Advancements made in genetically modified crops, such as soybeans resistant to the herbicide glyphosate, and the development of new fertilizers, insecticides, and herbicides have contributed to adoption by large and medium scale farmers.

Equipment and fuel costs are the greatest cost consideration for large scale farmers considering adopting CA. In the early 1990s lighter, precision seeding models were released that helped to spur adoption. In Brazil there are now 300 different models of commercial no-till seeders (Calegari, et al., 2013). While the tractors used for no-till planting do not require as much power or fuel, reducing those costs, the switch to CA does require an initial investment in new planting equipment. Analyses show that labor, fuel, and equipment costs are smaller over time, but additional costs such as herbicide and pest management can offset these savings. Comparison of CA systems with conventional tillage systems has not shown consistently higher financial returns except where erosion was an issue degrading soil fertility and where shorter field preparation time with CA enabled more crops or double cropping during a growing period.

Over the long term, CA has been shown to increase profitability over conventional farms in some areas, as a 10-year comparison of 18 large and medium-sized farms in Paraguay showed a 300% increase in net income while the net income on conventional farms fell during that same period (Sorrenson, 1997). In the US, greater yields were found with adoption of no-till in the western and southern regions where the water conserving aspects of CA helped address water shortages, however in northern regions, especially on poorly drained soils, yields could be lower where no-till postponed planting dates and short-ened the total growing season with soils that stayed cooler longer in the spring.

No-till planting equipment has evolved to increase efficiency and yield with more precise seeding. To increase the effectiveness of planters on heavy clay soils, double disc openers were replaced with a ripper tine on some planting equipment. Seeding speeds can result in bouncing and misplaced seeds. Equipment manufacturers develop solutions to these and other problems with devices such as rebounder attachments and other innovations. Precise seed spacing is critical to reduce seed gaps, weed pressure, and competition between plants, and can be accomplished with metering systems. Seed firmers have been developed to increase uniformity in seed depth. Special closing wheels have also been added to cover seeds with soil that is less dense.

Because residue can interfere with planters and seed placement, and can be large for some crops such as corn, double disc openers have been developed to slice through residue to improve seed placement. Some combines used for harvesting also have implements that help to chop and evenly spread the residue. Strip headers have been developed to strip the grain from the stalk and some combines have residue shredders that shred residues as they pass through the combine. Recent equipment innovations have increased speeds of planting, to increase the amount of land that can be planted per day for larger farms.

Some farmers both in tillage agriculture and no-till are adopting precision agriculture with geographic information systems (GIS) and in-field sensors that can provide feedback about the needs of the crop and problems in the field. This data assists with optimal timing of planting, fertilization and harvest to improve yield and profits and lower costs. Equipment manufacturers and fertilizer companies have also come forward with interactive decision support tools for precision agriculture and other services to assist growers.

In many developed countries, especially in Europe, reduced and no-till has not been adopted in large part due to challenges managing weeds without tillage, especially where government policy restricts elevated use of pesticides and herbicides. Some research has explored alternatives to herbicidal control of weeds to address the needs of no-till and organic agriculture and in considering the advent of herbicidal resistant weeds. Research in Brazil has shown that some cover crops such as hairy vetch (*Vicia villosa* Roth), black oat (*Avena strigosa* Schreb.), and oil seed radish (*Raphanus sativus* L.) are effective at reducing weed populations and reducing the amount of herbicide needed. Methods such as increasing cover crop biomass to suppress weeds and allelopathic interactions have been studied with mixed results.

Trends in Hoe-Till Adoption in Subsistence Agriculture

Most progress in adoption by smallholder subsistence farmers has been in South America; adoption has been marginal outside of Brazil, Paraguay, and Uruguay, where government programs have provided support and education through extension services to promote adoption by small farmers. Studies of net farm income of smallholders in South America were greater with CA than conventional practices. Animal driven rippers and seeders were developed in Brazil for small scale farmers, and have been exported to Africa for smallholder farmer mechanization.

Other "homemade" equipment developed by and for small farmers includes the knife roller (roller-crimper) that is designed to bend over and crush cover crops and weeds, and other vegetation flatteners to press the residue down before planting. Wooden and metal subsoilers where the moldboard plow share is replaced with a metal point to reduce soil disturbance have been developed that can be used with draft animals. Also a metal chisel based on the conventional subsoiler design have replaced the metal point with a thinner and longer metal spine developed to pierce through compacted soil layers while disturbing the soil less.

Weed control becomes an issue if there is not enough labor or no access to herbicides. There is also a common view that CA requires increased fertility, which in the past has been accomplished by plowing residues into the soil. Without that flush of nutrients, fertilizer input at least in the initial stages would be required to avoid a reduction in yield with CA. Fertilizer inputs for the smallholder farmer are often expensive or not accessible.

Many smallholder farmers in Africa must consider the use of residues as feed and bedding for livestock where they are integrated into farming. Residues have other purposes such as construction and as a fuel source. Smallholder farmers also face issues in adopting crop rotations, when there are no markets for alternative crops, especially legumes.

Intercropping of legumes may be one solution to increase residue production and nutrient requirements. However, the timing and crop selection are important to avoid competition between crop plants and lower yields.

With respect to equipment in South Asia, two-wheeled tractor-mounted planters are replacing animal drawn plows and planters.

Summary of CA Benefits

By increasing biodiversity, increasing water infiltration, and reducing soil erosion, CA has multiple benefits for agriculture and ecosystems. In summary, CA:

- reduces soil erosion and degradation, which:
    - maintains soil structure, quality, and fertility
    - improves water and air quality
    - reduces vegetation/ecosystem loss
    - reduces loss of nutrients
    - reduces sediment buildup in reservoirs
- increases water infiltration, which:
    - enhances resilience to drought and other extreme weather, such as hurricanes
    - reduces surface runoff
    - reduces flooding
    - reduces soil erosion
    - recharges groundwater and aquifers
- increases soil organic matter, which:
    - improves soil nutrients and water storage capacity
    - improves soil fertility and structure
    - reduces dependence on chemical inputs reducing environmental degradation
    - restores agroecosystem health
    - sequesters carbon
- maintains crop yields and often increases yields on soils that have been degraded, which:
    - increases food security and reduces rural poverty,
    - minimizes expansion of agriculture into less suitable areas, reducing deforestation and native ecosystem destruction
- reduces costs of farming, by reducing fuel use, equipment maintenance, and can reduce labor and production costs over time, which:
    - increases profits for economic sustainability
    - reduces greenhouse gas emissions from tillage and replacing tillage equipment (tilling the soil consumes more energy than any other farming operation)
- Enhances natural resources and biodiversity of agroecosystems, which:
    - increases the variety of soil organisms, including both fauna and flora, including wildlife.

Bibliography

Calegari, A., Guilherme de Araujo, A., Costa, A., Lanillo, R. F., Casao Junior, R., & Rheinheimer dos Santos, D. (2013). In R. A. Jat, K. L. Sahrawat, & A. H. Kassam, *Conservation Agriculture in Brazil of Book Conservation Agriculture: Global Prospects and Challenges*, (Chapter 3, p. 85. Oxfordshire, UK: CABI.

Darwin, C. (1881). *The Formation of Vegetable Mould, Through the Action of Worms, With Observations on Their Habits*. London: John Murray.

United Nations (2015). "What is Conservation Agriculture?" Retrieved from: Food and Agriculture Organization of The United Nations: Helping to build a world without hunger. Available at: www.fao.org/ag/ca/1a.html (accessed on February 9, 2015).

FAOSTAT. (2014). AQUASTAT database – UN FAO. Available at: www.fao.org/nr/water/aquastat/data/query/index.html (accessed on July 16, 2014).

Faulkner, E. H. (1943). *Plowman's Folly*. London: Michael Joseph Ltd.

Fukuoka, M. (1978). *The One Straw Revolution: An Introduction to Natural Farming*. Emmaus, Pennsylvania: Rodale Press.

Pope, R. O. (1989). Age-hardening behavior of two Iowa soil materials. Doctoral dissertation. Iowa State University.

Sorrenson, W., Portillo, J. L., Derpsch, R., & Nunez, M. (1997). Economics of no-tillage and crop rotations compared to conventional cultivation cropping systems in Paraguay. *Bibliotheca Fragmenta Agronomica*, 2.

CHAPTER 12

Soil Classification and Surveys

Soil classification makes safe and productive uses possible for each kind of soil. In this time of increasing pressure on the land, the systematic approach of modern soil classification is a great help in avoiding abuse of soils and mistaken investments in land and operations that are incompatible with soil conditions. Soil classification and mapping (surveys) permit the transfer of soils information from one place to another and from present to future generations.

Scientific soil classification is generally recognized to have begun in 1885 when the Russian scientist V.V. Dokuchaev undertook the study and classification of soils near Moscow. At about the same time, in the United States, similar concepts of soils as natural bodies in the landscape were being formulated by E.W. Hilgard. C.F. Marbut, the director of the U.S. Soil Survey Division for the first 35 years of the twentieth century, introduced many of the Russian concepts of soil science. Marbut's successor was C.E. Kellogg, who was the primary author of the first U.S. system of soil classification that was published in the U.S. Department of Agriculture (USDA) *1938 Yearbook of Agriculture.*

It soon became clear that this system was inadequate, in part because it did not incorporate specific boundaries on soil properties in order for soils to be classified within a certain group. As a result, the soil survey staff began work on a modern system of soil classification in the 1950s. The product of their investigations and deliberations, under the leadership of Guy Smith, went through successive approximations, and in 1960 *Soil Classification, 7th Approximation* was published.

Taxonomy is the science of classification. The specific subdivision of soil science that deals with soil classification is **pedology,** and those who specialize in soil classification are **pedologists.**

After much worldwide scrutiny and many amendments, the 1975 edition of *Soil Taxonomy* was published. At that time, 10 orders were recognized. The refinements continued, and in 1998 the current revision of *Soil Taxonomy,* with 12 soil orders, became available.

Soil Science Simplified, Sixth Edition. Neal S. Eash, Thomas J. Sauer, Deb O'Dell, and Evah Odoi.
© 2016 John Wiley & Sons, Inc. Published 2016 by John Wiley & Sons, Inc.

The Soil Classification Categories

The soil classification system of the USDA is hierarchical with these six categories:

Order
    Suborder
        Great group
            Subgroup
                Family
                    Series

The highest category, soil order, is the most generalized. As an individual soil is classified down through the system to the lowest category, soil series, an increasing number of specific properties are recognized. This system attempts to precisely categorize soils over the entire face of the earth in one of the 12 soil orders. It should be recognized that there are extensive land areas that are not soil. These include rocky land, shifting sand, and ice/glaciers.

The 12 Soil Orders

In this book, the soil orders are grouped according to natural characteristics based on the five soil-forming factors in an effort to make them easier to remember. These factors are parent material, climate, living organisms, topography, and time.

Each soil order has a formative element that is a descriptive term consisting of two to three letters that are set in italics. These letters form the last syllable of the names in lower categories in that order. After each soil order heading, the estimated percentage of the world's soil in that soil order is given. Table 12.1 gives a summary of the 12 soil orders, their characteristics, and approximate classification in the Food and Agriculture Organization of the United States (FAO) system that is used in many parts of the world. A world map of the 12 soil orders based upon *Soil Taxonomy* is presented as Figure 12.1.

Time Is too Short for Strong Soil Development

Entisols: Soils Having Minimal Development (16.2%)

The order Entisol (the root word is the English word "rec**ent**") includes soils that are so weakly developed they may have only a thin ochric epipedon over a C horizon (Table 12.1, Fig. 12.2). There are two reasons why these soils lack greater development: (1) The parent material consists of such highly resistant minerals that the rate of weathering is very slow; for example, droughty sands remain poorly developed because they contain an abundance of quartz; or (2) The exposed land surface is young as a result of erosion or burial under new material brought in by wind or some other agent. Unprotected soils on slopes are subject to water erosion, and on more-level plains wind may erode the topsoil, thus exposing new material below and keeping the soil young. The opposite process keeps alluvial and wetland soils young. New material is added layer by layer as floodwater moves over the soils or temporary ponds spread across them.

Table 12.1 The 12 soil orders used in *Soil Taxonomy*,[1] their formative elements, correlating FAO classification, and U.S. and worldwide distribution

Formative elements	Resulting soil orders	Description	Common master horizons	Common subsurface horizons	Description of genetic horizons	FAO translation	% of U.S. land area[1]	% of world land area
Time	Entisols	Young, thin sola, relatively underdeveloped soils	A C	None	Thin A horizon over parent material	Fluvisols Lithosols Arenosols Regosols	16.2	17.9
	Inceptisols	Older than Entisols	Ap Bw C	Ap (plowed) Bw (cambic)	Weak B horizon	Gleysols Solonchaks Rankers Cambisols	9.8	15.2
	Oxisols	Very old, thick sola, highly weathered	Ap Bo (several) Bt C	Ap (plowed) Bo (oxic) Bt (argillic) By	Weathered B horizon, weak Argillic	Ferralsols	7.5	0.02
Climate	Aridisols	Arid, can be highly developed	A Bk, Btk, By, or Bz Btk C	Bt (argillic) Bz (accumulation of salts more soluble than gypsum) By (gyspsic)	Due to limited leaching, accumulation of various salts	Xerosols Yermosols Solonchaks	12.0	11.8
	Spodosols	Formed under acid forests	Oie A E Bhs C	E (eluvial, if very light, Albic horizon) Bh (illuvial organic matter) Bs (illuvial sesquioxides)	Acid litter causes eluviation of many compounds	Podzols	2.7	3.5
	Gelisols	Permafrost				Gleysols Regosols	8.6	9.1
Topography	Histosols	Organic, wetland soils	Oap Oa Oe	The O horizons are organic materials (Oa are highly decomposed, Oe are of intermediate decomposition, Oi are slightly decomposed)	Differentiated based upon degree of decomposition of organic matter	Histosols	1.6	1.2

(continued)

Table 12.1 (*Continued*)

Formative elements	Resulting soil orders	Description	Common master horizons	Common subsurface horizons	Description of genetic horizons	FAO translation	% of U.S. land area[1]	% of world land area[1]
Biota	Mollisols	Dark soils formed under prairies, high base saturation	Ap Bw,k,t,y,orz C	Ap (plowed) Bt (argillic) Bk (carbonates) Bn (sodium)	Dark A horizon, B horizon can have a variety of endopedons	Gleysols Solonchaks Andosols Rendzinas Planosols Solonetz Kastanozems Chernozems Phaeozems Greyzems Podzoluvisols	21.4	7.0
	Alfisols	Brown soils formed under deciduous forest, high base saturation	Ap E Bt C	Ap (plowed) Bt (argillic) Bx (fragipan)	Thin, brown colored A horizon, must have an argillic horizon	Solonetz Luvisols Planosols Nitosols	13.9	10.1
	Ultisols	Brown soils formed under deciduous forest, low base saturation	Ap E Bt C	Ap (plowed) Bt (argillic) Bx (fragipan)	Thin, brown A horizon, must have an argillic horizon and/or a kandic horizon	Acrisols Nitosols Planosols	8.5	8.1
Parent material	Andisols	Formed from volcanic ash, young soils	Ap Bw C	Ap (plowed) Bw (cambic) Bh (humus)	B horizons are commonly weakly developed	Andosols	0.8	1.7
	Vertisols	Have expanding clays	Ap Bw C	Ap (plowed) B horizons are weakly developed due to constant churning Due to clay expansion and contraction	Self-churning soils due to expansion and contraction of clays.	Vertisols	2.0	2.4

[1]Natural Resources Conservation Service. Land use areas are approximate and do not sum to 100% due to unknown factors.

12.1 World soil map.

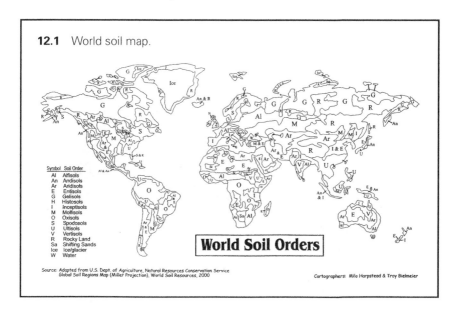

World Soil Orders

Source: Adapted from U.S. Dept. of Agriculture, Natural Resources Conservation Service
Global Soil Regions Map (Miller Projection), World Soil Resources, 2000

Cartographers: Milo Harpstead & Troy Bielmeier

Symbol	Soil Order
Al	Alfisols
An	Andisols
Ar	Aridisols
E	Entisols
G	Gelisols
H	Histosols
I	Inceptisols
M	Mollisols
O	Oxisols
S	Spodosols
U	Ultisols
V	Vertisols
R	Rocky Land
Sa	Shifting Sands
Ice	Ice/glacier
W	Water

12.2 Entisols are weakly developed.

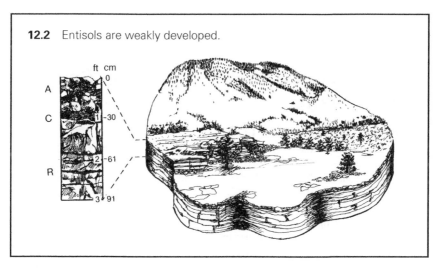

Inceptisols: Immature Soils (9.8%)

Inceptisols (from the Latin word *inceptum* which means "beginning") show more development than Entisols, but compared to other soils in the same region, they are immature (Fig. 12.3). They are found in most climatic zones but are excluded from arid regions and where there is permafrost. Most Inceptisols have an ochric epipedon and a cambic B horizon.

Sloping mountainsides are commonly occupied by Inceptisols because geologic erosion and leaching by rainfall is ineffective at such sites. Inceptisols are also common in depressions. One reason is that soil development is slowed where the soil does not dry out periodically.

12.3 Inceptisols are relatively immature.

Climate Is the Dominant Factor in Soil Development

Aridisols: Desert Soils (12.0%)

Aridisols (from the Latin word *aridus* which means "dry") are desert soils that are dry nearly all the year (Fig. 12.4). They have an ochric epipedon and may have either a cambic or an argillic subsurface (B) horizon. Entisols are commonly mapped near Aridisols. Some, but not all, desert soils are salty. Aridisols often contain calcium carbonate that can result in a calcic subsoil horizon called a calcic horizon. Petrocalcic horizons are cemented calcic horizons that are impenetrable by roots. Aridisols often have silica-cemented hardpans called duripans that are rock-like and also impenetrable by roots.

The deserts are very fragile regions, and once they are disturbed they are slow to recover. For this reason, intensive recreational use of the desert is viewed with alarm by environmentalists. Early records indicate that many Aridisol regions were once quite grassy whereas now only scattered shrubs grow as a result of overgrazing and other uses.

Gelisols: Always Frozen Soils (8.6%)

The central concept of Gelisols (from the Latin word *gelare* which means "freeze") is that they contain gelic material underlain by permafrost within 40 in. (100 cm) of the soil surface (Fig. 12.5). Gelic materials are mineral and/or organic matter that has been mixed in various patterns due to the churning caused by freezing and thawing in the active layer above the permafrost. They usually also exhibit ice segregation in this layer. Gelisols are found extensively in Alaska, Canada, Greenland, Iceland, and Siberia.

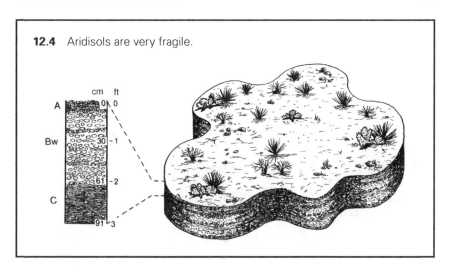

12.4 Aridisols are very fragile.

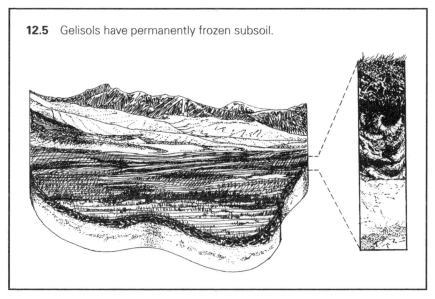

12.5 Gelisols have permanently frozen subsoil.

Oxisols: Highly Weathered Tropical Soils (7.5%)

Oxisols (from the French word **ox**ide) are highly weathered soils and form most commonly from sedimentary rocks and basic crystalline rocks that are relatively susceptible to weathering. An ochric epipedon overlies an oxic subsurface diagnostic horizon. Oxisols develop in tropical areas (Fig. 12.6) and have a high content of inert clays, mostly amorphous oxides of iron and aluminum. The only kind of layered silicate clay found in more than trace amounts is kaolinite. Ironstone nodules, which may contain considerable amounts of manganese, are sometimes present in the soil profile.

Oxisols usually have a granular structure throughout, which allows them to absorb water readily and makes them easy to till. Nutrients are quickly lost from Oxisols when

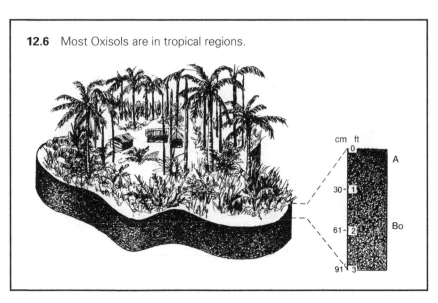

12.6 Most Oxisols are in tropical regions.

they are tilled because their humus decomposes quickly and their clays have a very low cation exchange capacity. Historically, farming has been carried out by a system of shifting cultivation, wherein the land is left to grow trees and shrubs for several years. This allows natural incorporation of nutrients absorbed by trees within the soil into the leaves of the trees and subsequently become organic residue at or near the surface. Clearing, burning, and a few years of cropping follows while the humus decomposes. This is called slash and burn agriculture. When crop yields decline, the cycle is repeated. Many Oxisol areas are used successfully for permanent crops such as cocoa beans and oil palm. Some areas of Oxisols are planted regularly to sugarcane, pineapple, and other tropical crops with the help of modern agricultural techniques.

Parent Material Is Specific

Andisols: Volcanic Soils (0.8%)

Andisols (from the Japanese word **and**o used to describe dark soil) are volcanic soils. Andisols formed from volcanic ejecta (pumice, cinders, lava) and closely associated parent materials in hilly or mountainous areas (Fig. 12.7). Fresh volcanic ash would not qualify as an Andisol and geologically old deposits grade into other soil orders. Andisols have early weathering products in the colloidal fraction (very small clay and humus particles), namely, allophane, imogolite, ferrihydrite, and aluminum-humus complexes.

Due to the resistance to decomposition of the metal-humus complexes, the organic matter commonly reaches 10–20%. This gives a melanic epipedon that has a very dark color. Andisols have a low bulk density that promotes water infiltration, low water erosion potential, and ease of tillage. They are noted for their high natural fertility and are often the most productive soils in their region. Crops such as coffee are grown extensively on them. Nevertheless, production may be inhibited by the tie-up of phosphate on their anion exchange complex.

12.7   Andisols have many layers of volcanic ash.

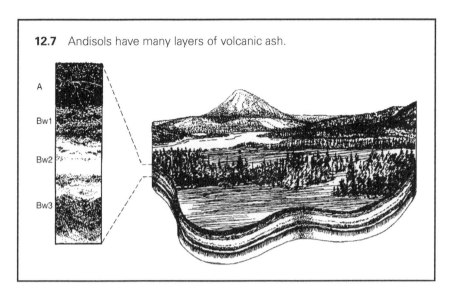

12.8   Histosols are accumulations of organic matter.

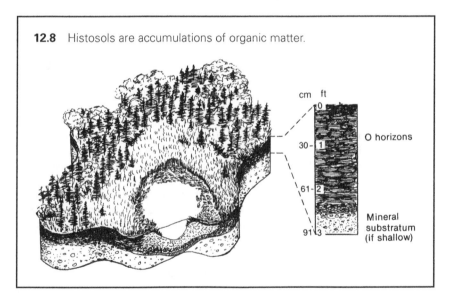

Histosols: Organic Soils (1.2%)

Histosols (from the Greek word *histos* which means "tissue") are organic soils that formed from accumulations of organic matter (Fig. 12.8) in wet/cool environments that slowed residue decomposition. Many Histosols formed in shallow lakes or tidal flats that accumulated plant residues. These soils act like a sponge and are saturated most of the time. Histosols on cool mountain slopes have a folistic epipedon and drain freely.

Poorly decomposed Histosols are commonly called peat and are not good for farming, even if they are drained. They are, however, sometimes harvested for use in greenhouses

12.9 One author stands beside a subsidence post at Belle Glade, Florida.

and nurseries. Well-decomposed Histosols are called muck and are often drained for specialized farming (vegetables, for example). The drainage and tillage of Histosols causes the pores to fill with air, which speeds decomposition resulting in subsidence at a rate of about 1-ft (0.3-m) drop of the surface every 10 years. The Everglades Agricultural Area in southern Florida is an example of an area with high agricultural productivity due to the successful draining of the Histosols. Unfortunately, this practice is not sustainable due to subsidence. Figure 12.9 shows the subsidence post at Belle Glade, FL. In 1924 the top of this post was even with the soil surface; in 2002 the top of the post is nearly 6 ft out of the soil surface. This Histosol formed in nearly 9 ft of organic material above the bedrock. In these places the soil must now be moved into ridges to have sufficient soil depth to plant crops.

Vertisols: Cracking Dark Clay Soils (2.4%)

Vertisols (from the Latin word *verto* which means to "turn") are dark clay soils formed most extensively in the warm temperate and tropical areas with an ustic moisture regime, but they may also be found in cooler climates (Fig. 12.10). For example, in the United States, most Vertisols have been mapped in Texas and South Dakota.

Vertisols owe their unique properties to the shrinking and swelling of clays. Wide cracks open during the dry season and some soil is likely to fall into them. When the rains return the cracks swell shut, but if they have been partially filled, there is not enough room for the cracks to close. This causes a churning action that brings up fresh limy material from the C horizon and thereby rejuvenates the topsoil faster than it can be leached by rainwater.

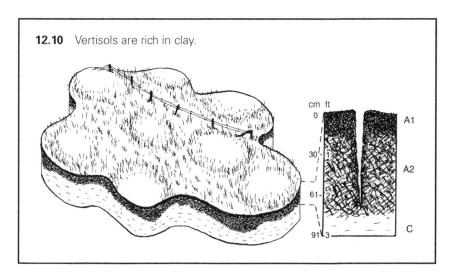

12.10   Vertisols are rich in clay.

The swelling action in Vertisols also causes lens-shaped blocks within the soil to slide past each other and develop polished surfaces called slickensides. The same forces buckle the landscape into mounds and hollows, which results in a microtopography called gilgai.

Vertisols are particularly well adapted to sugarcane and paddy rice culture. In Texas, Vertisols are largely used for pasture despite the fact that in dry seasons open cracks make footing hazardous for cattle. Within a few years, fence posts, telephone poles, and buildings may become tipped and twisted on these soils. When houses are built on them, the builder may provide a way to keep the soils under the foundation moist at all times to prevent the heaving action that can ruin a building.

Vegetation Is a Grassland (Prairie)
Mollisols: Grassland Soils (6.9%)

Mollisols (from the Latin word *mollis* which means "soft") are grassland soils. They are among the most productive agricultural areas of the world (Fig. 12.11). In the United States most Mollisols are found in the central northern prairie states that comprise the Corn and Wheat Belts. The general properties of soils developed under prairie vegetation were discussed earlier in this chapter, and nowhere are they more strongly reflected than in the Mollisols. The dense fibrous root system of the grasses and forbs has resulted in the development of a thick, dark, humus-enriched A horizon (mollic epipedon) with an abundance of plant nutrients. These soils were first described in Russia, where the darkest were called Chernozems, meaning black earth.

In the humid part of the Mollisol area in the midcontinental United States— the Corn Belt—the subsoil has an accumulation of clay (argillic horizon). This property is minimal or even absent in drier or colder parts of the grasslands where a cambic diagnostic subsurface horizon prevails. In North America, Mollisols are most extensive on the tall- and short-grass prairies that extend eastward from the Rocky Mountains. Many settlers built their first houses of sod in which soil was bound together by grass roots.

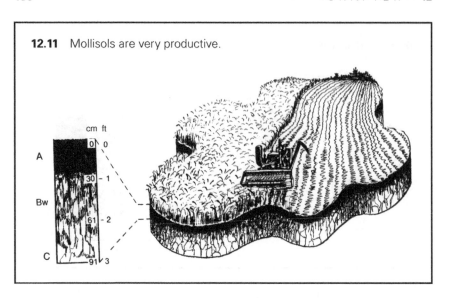

12.11 Mollisols are very productive.

Climate and Vegetation Combination Dominates

Alfisols: High-Base Status Soils of Hardwood Forests (9.7%)

Alfisols (from the term "pedalfer" in 1938 system) are typically found under deciduous forests where recycling of plant nutrients is effective (Fig. 12.12). Calcium, for example, is absorbed by plants and moved into the leaves and is returned to the soil when the leaves fall. Normally, the parent material contains calcium carbonate and is medium to fine textured. The humus-enriched A horizon is not thick and is therefore called an ochric epipedon. The subsoil has an accumulation of clay with dark clay films on the structural (ped) surfaces. This is the argillic horizon, which holds moisture and nutrients within the upper part of the root zone and is, therefore, beneficial to plants.

In the parts of the tropics where geologic erosion prevents the land surface from becoming highly weathered, the soils may grade from Inceptisols to Alfisols with the increasing stability of the landscape. When Alfisols are cleared of their timber and placed under cultivation, they are usually quite productive and respond well to fertilization. The good supply of water, timber, and agricultural land in Alfisol regions throughout the world accounts for the development of large centers of population on them.

Ultisols: Low-Base Status Forest Soils of Warm Regions (8.5%)

Although cycling of bases (calcium, magnesium, potassium) goes on in Ultisols (from the Latin word *ultimus* which means "last") under forest cover, it is less effective than in Alfisols because the geologic substratum usually lacks calcium carbonate. Leaching, which occurs year-round, has removed many plant nutrients from the root zone. There is a definite accumulation of clay in the subsoil (the argillic horizon), but it is highly weathered.

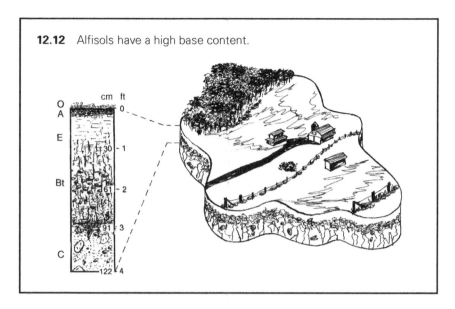

12.12 Alfisols have a high base content.

In Ultisols the ultimate weathering of layered silicate clays has taken place. As a result, kaolinite is abundant. These soils are usually considered older by tens of thousands of years than the Alfisols. Well-drained Ultisols are brightly colored by stains of yellow and red iron oxides. Poorly drained Ultisols are gray. If Ultisols are cultivated, they quickly become impoverished unless fertilization and careful management are practiced (Fig. 12.13).

Historically, in the southern United States, many Ultisols were planted to cotton which have very low yields after a few years of production. Often severe erosion followed and sometimes abandonment of the land, which continued to erode. By use of modern methods, many of these farms are being brought back into production; many even yield two crops per year due to the long growing season. In tropical areas three crops per year are possible with intensive management.

Vegetation and Parent Material Dominate

Spodosols: Acid Soils of Sandy Pine Lands (2.7%)

Spodosols (from the Greek word *spodos* which means "wood ash") are most common in the sandy outwash regions of the boreal forests and in quartzose (sandy) coastal marine deposits extending to the tropics. Under the acid humus layer (Fig. 12.14), there is likely to be a whitish albic horizon overlaying a dark brown spodic horizon in which humus and or iron oxides coat the sand grains. Sometimes these coatings cement this horizon into a pan called an ortstein. The village of White Earth, Minnesota, was named for this white soil horizon. Much of the pine lumber that went to build towns and farmsteads throughout North America and Europe came from the extensive forests of the Spodosol regions. However, when the forests were cleared, these soils, which had yielded such beautiful timber, did not prove to be good for base-loving, shallow-rooted agricultural crops. Many farms failed and were replaced by pine plantations and mixed forests for lumber and pulp production as well as for use by wildlife and for recreation.

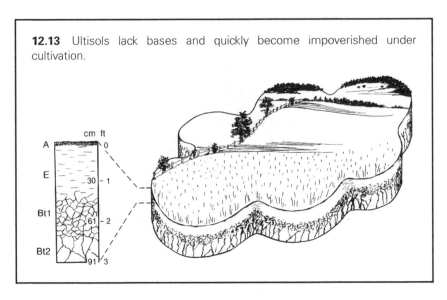

12.13 Ultisols lack bases and quickly become impoverished under cultivation.

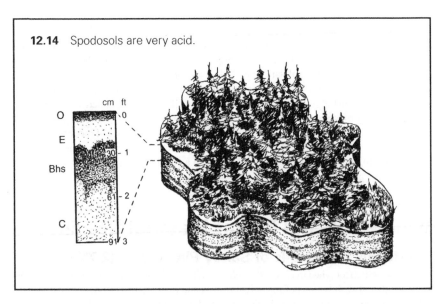

12.14 Spodosols are very acid.

Lower Categories of the Classification System

As was stated earlier, there are six categories in the USDA soil classification system. They are order, suborder, great group, subgroup, family, and series. Each of these categories is regulated by the Natural Resources Conservation Service of the USDA. The first five categories are defined in *Soil Taxonomy,* and individual sheets are published for each soil series. In addition, two other categories, type and phase, may be described locally for land-use planning purposes. With each successive category, more information is revealed about the soil being classified.

In the following paragraphs, each category below the order of the soil classification system is explained and examples are given.

Suborders

The suborder category uses a formative element (syllable) from the order name and places a new syllable before it to give more information about the soil. In many cases the new syllable may indicate such features as the usual moisture condition and a particular property of the parent material, or as in the case of Histosols, the degree of decomposition of organic materials. For example, the "oll" from Mollisol is used to identify the soil order in soil classification (Table 12.1).

"Oll" is joined with another syllable such as "ud" (from *udic*) to make the two-syllable "udoll." A udoll is a suborder common in humid regions (common in the Corn Belt) whereas an ustoll would be intermediate in moisture between a humid and arid climate, soils common in the Wheat Belt states. Similarly, an "aquoll" is a wet soil where the water table will be near the soil surface sometime during the year. Most aquolls require drainage before they can be intensively cropped. Some syllables are used at more than one level in the soil classification system. Table 12.2 illustrates the process for classifying soil profiles.

Great Groups

Great groups, the third category, are formulated by adding another syllable in front of the suborder name to give more information about the soil properties. Five examples of these are given below:

Arg—Latin *argilla*, white clay
Ust—Latin *ustus*, burnt (dry climate)
Hapl—Greek *haplous*, simple
Gloss—Greek *glossa*, tongue
Pale—Latin *paleos*, old

Sometimes a vowel is placed between these syllables and the suborder name to make the word easier to say:

Argiudoll—a Udoll with clay accumulation in the B horizon
Ustipsamment—a Psamment of dry regions
Haplorthod—a simple Orthod
Glossocryalf—a Cryalf with the E horizon irregularly protruding into the B horizon
Paleaquult—a highly developed Aquult evidenced by genetic horizon thickness or number

Subgroups

The subgroups are formed by modifying the great group name with one or two adjectives. This adjective may depict a normal condition, or it may indicate some special features about a soil. In some cases a great group of one order may be integrated toward another, and this would be shown by the subgroup adjective. Five examples are given below:

Typic—fits the central concept of the great group
Aquic—having properties of wetness
Alfic—grading toward an Alfisol (with an argillic horizon)
Fragic—having a fragipan
Aeric—periodic aeration

Table 12.2 *Soil Taxonomy* classification scheme

Order	Suborder	Great group	Subgroup	Family	Series	Description	Areas where commonly found
Gelisols	*Turbels*	*Histo*turbels	*Typic* Histoturbels	*Loamy-skeletal, mixed, superactive, subgelic shallow* Typic Histoturbels	*Ester*	Permafrost within 1 m of soil surface Profiles with cryoturbation More than 30% organic materials to a depth of 0.5 m Fit within the concept of a Histoturbel Has loam texture, rocks within the profile, mixed mineralogy, high-activity clays, cold, shallow Poorly drained soil formed in thin loess over schist bedrock	Alaska
Histosols	*Fibrists*	*Cryof*ibrists	*Lithic* Cryofolists	*Dysic Typic* Cryofolists	*Reggad*	Organic soil Organic fibric materials that are not identifiable Mean temperature is <8°C, no permafrost Bedrock at <0.5 m Acidic Deep, excessively drained soil formed in decomposing organic matter with a mixture of ash and pumice over rubble	Washington
Spodosols	*Cryods*	*Humicryods*	*Andic* Humicryods	*Sandy-skeletal, mixed* Andic Humicryods	*Chugach*	Have a spodic horizon Mean temperature is <8°C, no permafrost Have >6% organic C in spodic Formed from volcanic ash and have andic properties Sandy texture with rocks in the profile Very deep, well drained soils formed in a thin mantle of ash overlying glacial outwash	Alaska

Order	Suborder	Great group	Subgroup	Family	Series	Description	Location
Andisols	*Torrands*	*Duritorrands*	Petrocalcic Duritorrands	*Medial-skeletal, amorphic isohyperthermic* Petrocalcic Duritorrands	*Hapuna*	Formed in volcanic materials; Soil profile is dry most of the year; Have a cemented horizon; Have a petrocalcic horizon; Soil formed in volcanic ash underlain by cobbles and lava; andic properties, temperature is >22°C throughout the year; Deep, well-drained soil that formed in volcanic ash	Hawaii
Oxisols	*Ustox*	*Eutrustox*	Kandustalfic Eustrustox	*Very fine, mixed isohyperthermic* Kandiustalfic Eutrustox	*Saipan*	Have an oxic horizon; Have an ustic soil moisture regime; Have high base saturation; Have low-activity clays, characteristics of an alfisol; Clayey, mixed mineralogy, temperature is >22°C throughout the year; Very deep, well drained soils formed from limestone	Hawaii
Vertisols	*Xererts*	*Calcixererts*	Typic Calcixererts	*Fine, smectitic, frigid* Typic Calcixererts	*Niter*	Formed in lacustrine deposits with expanding clays; Have a xeric soil moisture regime; Have a calcic horizon; Fit within the central concept of a Calcixerert; Silty clay texture in soil profile control section, 2:1 expanding clays, mean annual air temperature of 5°C.; Deep, well-drained soil formed in lacustrine.	Idaho

(continued)

Table 12.2 (Continued)

Order	Suborder	Great group	Subgroup	Family	Series	Description	Areas where commonly found
Aridisols						Have an aridic soil moisture regime (soil profile dry in all parts most of the year), has a salic horizon (salt accumulation)	
	Salids					Have a salic horizon	
		Haplosalids				Fits within the concept of the salids suborder	
			Typic Haplosalids			Fits within the concept of the Typic Haplosalids Great Group	
				Fine-silty, mixed, superactive, hyperthermic Typic Haplosalids		Silty clay texture, mixed mineralogy, high-activity clays, soil temperature >22°C during the summer	
					Yahana	Deep, well-drained saline–sodic soil that formed in flood plain alluvium	Arizona
Ultisols						Older soil with an argillic or kandic horizon or a fragipan and argillic or kandic horizons	
	Udults					Have a udic soil moisture regime	
		Fragiudults				Have a fragipan	
			Glossic Fragiudults			Have a gray matrix color between the kandic or argillic horizon and the fragipan	
				Fine-silty, siliceous, semiactive, thermic Glossic Fragiudults		Silt loam texture, silica-cemented fragipan, low-activity clays, mean annual temperature is between 15–22°C	
					Dickson	Deep, moderately well-drained soil formed in loess	Tennessee
Mollisol						Grassland soils	
	Udoll					Udic soil moisture regime	
		Argiudoll				Has an argillic horizon	
			Typic Argiudoll			Typical Mollisol with an argillic horizon	
				Fine, mixed, active, mesic Typic Argiudolls		Clayey, mixed mineralogy, mesic soil temperature	
					Woolper	Deep, well-drained soils on footslopes, alluvium, or colluvium parent materials	Corn and Wheat Belt

Order	Suborder	Great group	Subgroup	Family	Series	Description	Location
Alfisol	*Aqualf*	*Epiaqualf*	*Mollic* Epiaqualf	*Fine-loamy, mixed, superactive, mesic* Mollic Epiaqualfs		Deciduous forest soils Poorly drained Soil is wet from water draining onto the site Epipedon dark enough to be mollic but usually too thin Texture of the soil control section, mixed mineralogy, mesic soil temperature	East of the Missouri River
					Havana	Deep poorly drained soil formed in loess or loamy calcareous glacial till	Iowa
Inceptisols	*Udepts*	*Dystrudepts*	*Fluvaquentic* Dystrudepts	*Coarse-loamy, mixed, active, frigid* Fluvaquentic Dystrudepts		Has a cambic horizon Has a udic soil moisture regime An acidic udept Formed in floodwater alluvium Coarse-textured profile, mixed mineralogy, high-activity clays, mean soil profile temperature <8°C	
					Podunk	Very deep, moderately well-drained soil formed in alluvium	Maine
Entisols	*Psamments*	*Quartzipsamments*	*Spodic* Quartzipsamments	*Hyperthermic, uncoated* Spodic Quartzipsamments		No subsurface horizons Loamy fine sand or coarser More than 90% resistant (quartz) material Has a horizon cemented by aluminium and organic matter, with or without iron Soil temperature >22°C during the summer, no coatings	
					Orsino	Very deep, moderately well-drained soils formed in sandy marine or aeolian deposits	Florida

[1] Adapted from NRCS soil descriptions.
[2] Each level of classification is highlighted in italics and described.

These terms may modify great groups to form subgroups as follows:

Typic Argiudoll—a typical Mollisol with an argillic horizon in a humid climatic zone
Aquic Ustipsamment—a slightly wet (seasonally) sandy Entisol in a dry climatic zone
Alfic Haplorthod—a simple, ordinary Spodosol having an argillic horizon below the spodic horizon
Fragic Glossocryalf—a cold Alfisol with highly irregular horizon boundaries and a fragipan
Aeric Paleaquult—an old, wet Aquult with colors indicating periodic aeration

The reader is reminded that each of these syllables indicates specific soil properties, as presented in the USDA publication, *Soil Taxonomy.*

Family

The fifth category of the classification system is the soil family. This is not named with strange-sounding Greek and Latin syllables but instead has descriptive terms indicating such properties as particle size, mineralogy, cation exchange activity, and temperature regime. A common family name for many soils that are formed from glacial till in mid-America is coarse-loamy, mixed, superactive, mesic. A finer-textured, highly weathered soil in the southeastern United States might be in the fine-loamy, kaolinitic, subactive, thermic family.

Series

The soil series is the sixth category and is the name given to soils with very similar profiles. The name that is given is derived from the town or community where the soil was first officially described. More than 20,000 soil series have been described in the United States and many more in other parts of the world. One example of a series is the Fayette, which is named after a town in Iowa. This soil formed in deep loess for some distance on each side of the Mississippi River in Iowa, Minnesota, Illinois, and Wisconsin.

Type and Phase

Even though not categorized in the soil classification system, the type and phase are added to further define a soil. Soil type gives the texture of the tillage zone. Soil phase gives information about soil properties that affect land use; slope and stoniness are examples. The type and phase are not numbered below because they are not a part of the six-category system for classifying soils. However, type and phase are useful in land-use planning.

A summary is given below for the naming of a system using an Alfisol:

1. Order—Alfisol
2. Suborder—Udalf
3. Great group—Hapludalf
4. Subgroup—Typic Hapludalf
5. Family—Fine-silty, mixed superactive, mesic
6. Series—Fayette
      Type—Fayette silt loam
      Phase—Fayette silt loam, nearly level

Soil Horizons

Pedons and Polypedons

Soil horizons were described in Chapter 2 as the layers that form in the soil during the long period of soil development. When viewed in a two-dimensional cross-section, such as

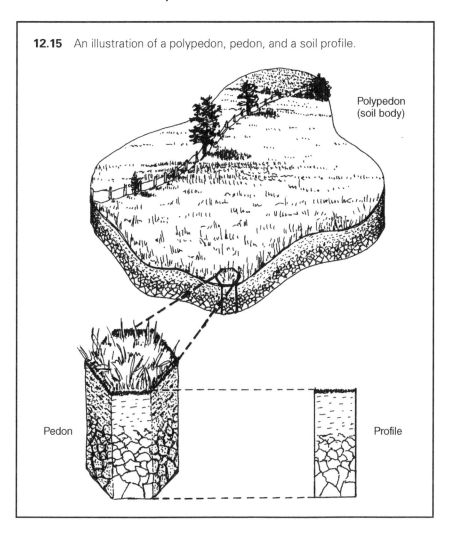

12.15 An illustration of a polypedon, pedon, and a soil profile.

Polypedon
(soil body)

Pedon

Profile

the side of a pit, they represent a soil profile. A soil profile extends from the ground surface to the depth of soil development.

The concept of the soil pedon considers the soil profile in three dimensions. The **pedon** is described as the smallest three-dimensional body of soil large enough to illustrate the nature and arrangement of soil horizons and their variability. The surface area of apedon is arbitrarily set at from 1 to 10 sq m, depending on the soil's uniformity or complexity. It is like a column of soil that would be left standing if a bulldozer removed all the soil except for that beneath a small patch of ground. Sometimes that is done if excavation takes place before the telephone company can reroute their lines.

A **polypedon** is defined as a set of contiguous pedons falling within the accepted range of characteristics for a specifically named soil on the landscape. The polypedon may also be termed a soil body. It is like having many columns of soil, side by side, extending to a boundary where a different kind of soil is encountered. The soil profile, pedon, and polypedon are illustrated in Figure 12.15.

Diagnostic Soil Horizons

Many types of soil horizons have been described by characteristics that fall within quantifiable physical and/or chemical parameters and meet a specified minimum thickness. At the highest level of soil classification, the soil order, there are three diagnostic surface horizons that are germane to soil classification. These named surface horizons may also be called epipedons. One of the epipedons, the mollic, has three named variations that will be discussed below.

Similarly, there are four diagnostic subsurface horizons used in the classification of most soils, at the order level, unless subsurface soil development is minimal. Various combinations of these seven diagnostic surface and subsurface horizons determine most soil orders. There are three major exceptions. The first is where extreme climate, very dry or very cold, makes traditional agriculture impractical. The second is where the soil parent material is volcanic ash. The third is where there is a thick accumulation of plant debris, such as in bogs and tidal flats.

Seven common diagnostic horizons for soil orders are as follows:

Diagnostic surface horizons (epipedons)
    Ochric (pale or thin topsoil)
    Mollic (thick dark topsoil, neutral to alkaline, fertile)
    Histic (thick organic mat over mineral subsoil)
Diagnostic subsurface horizons
    Cambic (only moderate soil development evident)
    Argillic (enriched in clay leached down from above)
    Spodic (enriched in colloidal humus, aluminum, and usually iron leached from surface horizons)
    Oxic (severely weathered, infertile, high in sesquioxide clays, usually reddish colored)

A few other diagnostic horizons that are variations of those listed above will be discussed with the major diagnostic horizons.

In most cases, the epipedons have an A horizon symbol, whereas the diagnostic subsurface horizons normally carry a B horizon symbol. Lower case letters are used to indicate more specifically the type of soil development that has occurred. Table 12.3 explains the use of symbols in soil horizon nomenclature.

Table 12.3   Examples of soil horizons

	Symbols		
Soil layer	General	Detailed	Properties
Solum			
Organic layer	O		
Leaf layer		Oi	Plant fiber recognizable
Humus		Oa	Saprophytes have decomposed the fibers
Topsoil	A		
Humus enriched		Ap	Darkened plowed layer
Subsoil			
Leached layer	E		Light colored due to fine particles being washed downward, eluviated
Accumulation zone	B	Bt	Where clay has moved in from above
Parent material	C		Little change by soil formation
Bedrock	R		The solid substratum

Description of the Diagnostic Surface Horizons (Epipedons)

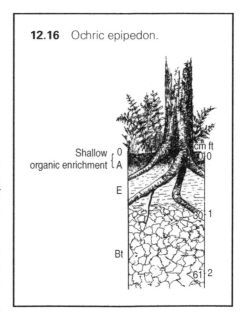

12.16 Ochric epipedon.

Ochric Epipedon

The ochric (from the Greek word *ochrose* which means "pale") epipedon (Fig. 12.16) is the most common type of A horizon. It may be a thin A horizon or one that is either pale or dark colored if it has <1.0% organic matter by weight. This is the usual condition where the native vegetation is a forest or the climate is arid. The average temperature conditions may be either warm or cold. When these soils are plowed, the fields have a grayish- or yellowish-brown appearance unless the farmer has added a lot of plant residue and manure to darken them.

Mollic and Similar Epipedons

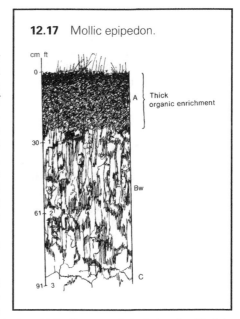

12.17 Mollic epipedon.

Mollic (from the Latin word *mollis* which means "soft") epipedons have a thick A horizon that is very dark brown or nearly black due to an enrichment of humus to >1.0% by weight (Fig. 12.17). This condition is usually met where the native vegetation is prairie grass. The grasses cycle basic ions to the surface, and the limited precipitation prevents rapid leaching and maintains a high base saturation. These soils develop a strong granular structure, which allows them to be crumbly even when they are dry.

Three variations of the mol-lic epipedon are as follows:

Umbric epipedons appear to be mollic, but they have a low base saturation due to their acidity. The name comes from the Latin word *umbra,* meaning shade, which alludes to their dark color. Umbric epipedons are not widespread, but humid or wet conditions prevail where they occur.

Melanic (from the Greek word *melanos* which means "black") epipedons are black, humus-enriched A horizons formed in loose volcanic materials. They have a low bulk density, are high in aluminum, and have a high phosphate retention capacity.

Anthropic (from the Greek word *Anthropoes* which means "human") **epipedons** are a human-induced form of the mollic epipedon. They most frequently occur in arid regions that have a long history of irrigated agriculture with organic matter incorporation. They are formed extensively in the Orient.

Histic Epipedon

A histic (from the Greek word *hist ose* which means "tissue") epipedon is an O horizon made up of plant residue 8–24 in. (20–60 cm) thick over mineral soil (Fig. 12.18). The range in thickness allowed depends on the degree of decomposition. These epipedons develop in lowlands that are saturated >30 days per year. Histic epipedons are subdivided according to their degree of decomposition. From the least to the greatest, the terms used are fibric, hemic, and sapric. Their notations are Oi, Oe, and Oa, respectively.

Folists (from the Latin *folia* which means "leaf") are a great group consisting of freely drained Histosols that are composed primarily of leaf litter and are underlain by fragmented bedrock or gravel within1mofthesoil surface. **Folist epipedons** are O horizons that can be as thin as 6 in. (15 cm) and are saturated for only a few days following heavy rains. They form in upland positions and are common in alpine regions where low temperatures cause plant residue decomposition to be slow.

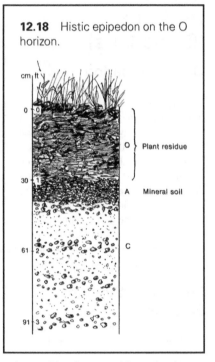

12.18   Histic epipedon on the O horizon.

Descriptions of Subsurface Horizons

Cambic Subsurface Horizon

When field and laboratory investigations of the subsoil show only a modest amount of weathering and not much accumulation of materials leached from above, the horizon is classified as a cambic horizon and has the symbol Bw. The word is derived from the Latin *cambire,* meaning to change. To qualify as cam-bic, the horizon must not be very sandy and must show some alteration by processes of weathering. These changes may be evidenced by changes in color, the development of soil structure, or the removal of some of its more soluble components.

In the subhumid parts of the Great Plains, soils usually have a cambic horizon below the mollic epipedon where prismatic structure has developed and from which carbonates have been leached. Figure 12.17 shows a Bw horizon of this kind. Figure 12.19 shows another type of cambic horizon where periodic wetness has brought about a mottled color due to the form and distribution of iron. It is indicated by the symbol Bg. The "g" is derived from the Russian word glei, meaning wet sticky clay, and the "w" indicates development of structure or color but little illuvial accumulation.

Argillic Subsurface Horizon

Throughout most of the humid hardwood forest region of the United States and some of the drier areas as well, the subsoils contain more clay than the A horizon and usually more than the C horizon. Some of the accumulated clay was moved down from the A and E horizons, and some was formed within the B horizon by the alteration of primary minerals into clay minerals. The small letter "t" in the horizon symbol Bt in Figure 12.20 is taken from the German word *tone,* meaning clay. Argillic horizons are subsurface diagnostic horizons where high-activity clays accumulated. The word is derived from the Latin *argillus,* meaning white clay. An argillic horizon usually benefits plants by holding moisture and nutrients within the root zone. To be classified as an argillic horizon, there must also be visual evidence of clay films lining pores or bridging sand grains.

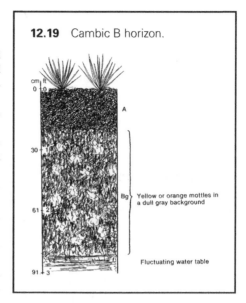

12.19 Cambic B horizon.

Natric Subsurface Horizon

A natric horizon (Btn) must have all the properties of the argillic horizon plus abundant sodium *(natrium* in Latin) that causes the soil to seal itself against the percolation of water. The impervious horizon illustrated in Figure 5.20 is a natric horizon.

Kandic Subsurface Horizon

Similar to the argillic horizon is the kandic (Bt) horizon found in subtropical and warmer regions. Kandic horizons have accumulations of low-activity clay, primarily kaolinite, and therefore do not hold nutrients well.

Spodic Subsurface Horizon

In boreal (northern) forest regions and some wet sandy areas of subtropical regions, the subsoil is usually reddish brown to black. This color is caused by coatings of humus together with iron and aluminum oxides on the surfaces of sand grains. These coatings may break off and become tiny pellets within the soil matrix. A subsoil layer with these properties is called a spodic horizon (Fig. 12.21).

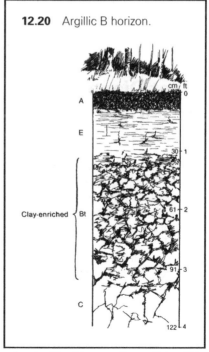

12.20 Argillic B horizon.

It may carry the symbol Bh, Bs, or Bhs; "h" represents humus and "s" represents iron and aluminum oxides (sesquioxides). When these soils are plowed, the light gray overlying albic E horizon associated with the spodic horizon gives the field an ashy appearance. The name comes from the Greek *spodos,* meaning "wood ash."

Oxic Subsurface Horizon

A very impoverished subsoil with almost no primary minerals other than quartz is called an oxic horizon (Fig. 12.22). It consists of quartz sand and an inert clay fraction of kaolinite plus oxides of iron and aluminum, hence, the symbol Bo. The entire subsoil is commonly quite uniformly weathered and lacks original rock features. Oxic horizons are found in tropical regions where severe weathering has been in progress for a very long time.

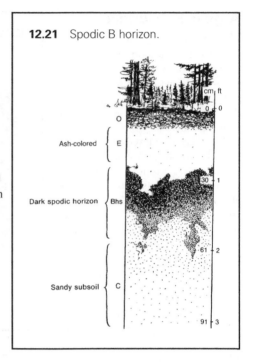

12.21   Spodic B horizon.

Other Diagnostic Subsurface Horizons

Albic Subsurface Horizon

Albic (from the Latin word *albus* which means "white") E horizons, similar to that shown in Figure 12.20, are whitish or gray in color with bleached, uncoated mineral grains. Usually they are below a mat of forest residue (O horizon) and at the surface of the mineral soil. The bleaching is, at least in part, due to organic acids leached from the O horizon. In this case, the albic horizon may be diagnostic for a soil order. Some may be shallow in the wet subsoil due to chemically reducing conditions.

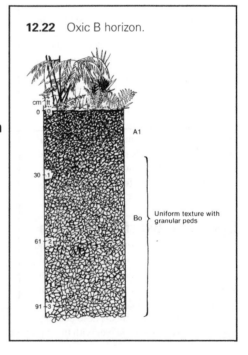

12.22   Oxic B horizon.

Calcic Subsurface Horizon

A calcic horizon (Bk) (from the German word *kalk* which means "lime") is an illuvial horizon in which secondary calcium carbonate or other carbonates have accumulated to a significant extent. They are widespread throughout grassland and desert regions.

Salic Subsurface Horizon

Salic horizons (Bz) (from the German word *zalt* which means "salt") are layers with a high accumulation of soluble salts, typically sodium chloride, in arid regions, where they may be diagnostic. The salt is derived from periodically shallow saline water in the subsoil.

Hardpans as Diagnostic Horizons

In many parts of the world, a subsoil hardpan exists. The layers have a very important effect on the potential use of the soil. They are not typically diagnostic at the order level, but they are recognized as diagnostic for lower categories in the soil classification system. Four prominent ones are described in this section. An example of a horizon notation symbol is given for each.

Petrocalcic (from the Greek word *petra* which means "rock") horizons (Bkm) occur in the subsoil on old landforms in arid regions where calcic horizon development has progressed to the point of becoming rocklike. It is composed mostly of calcium carbonate hardened around silicious gravel (Fig. 12.23). Petrocalcic horizons may extend to a depth of several feet, and are usually impenetrable by roots. The "m" indicates induration.

Duripans (from the Latin word *durus* which means "hard") are more durable than petrocalcic horizons because the cement in duripans includes much secondary silica (SiO_2). The symbol for them is Bq ("q" is from quartz). They form best where there is or has been volcanic ash and the climate has alternating dry and wet seasons.

Fragipans (from the Latin word *fragilus* which means "brittle") that occur in some forested regions are so dense they restrict the penetration of water and roots. The close

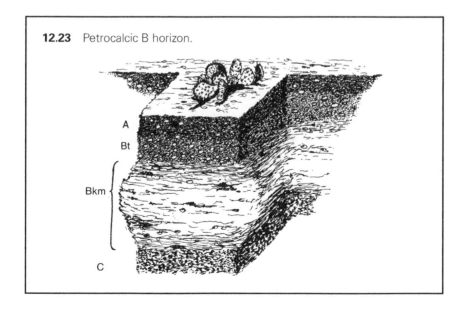

12.23   Petrocalcic B horizon.

packing of grains of sand and silt and weak cementation cause the horizon to be brittle when dry or moist, but not when wet. Most are Bx or Btx horizons.

Plinthite (from the Greek word *plintos* which means "brick") horizons (Bv) form in warm humid regions where iron is abundant in parent materials. The iron becomes concentrated and cemented into a continuous reticulate (netlike) layer. Plinthite may harden irreversibly into an iron pan (petroferric layer) when exposed repeatedly to wetting and drying over a long time.

Soil Moisture and Temperature Regimes

In the classification of soils, *Soil Taxonomy* takes into account not only soil pedon characteristics, but also soil moisture and temperature regimes. Moisture regimes relate to the water available to plants in the main part of the root zone. Temperature regimes are measured at a depth of 20 in. (50 cm) or to a restrictive layer if it is shallower. Each regime has established parameters, but in this book only general features will be presented. The terms below are incorporated into the names of many of the soil taxonomic units. The letters in bold are the "syllable" one finds in the taxonomic classification; those without bold letters indicate that these terms are used unabbreviated at the family level of soil taxonomy.

Classes of soil moisture regimes include the following:

Aquic—Saturated for enough of the time most years to cause reducing conditions (lack of oxygen) to prevail; soil moisture regime of wetlands.

Aridic and **torr**ic—Both terms are used to indicate dryness that restricts crop production without irrigation; soil moisture regime of the deserts.

Udic—Usually moist, soil moisture regime of the Corn Belt of the Midwest.

Ustic—Seasonal dry periods, but enough precipitation during the growing season most years for crop production without irrigation, soil moisture regime of the Wheat Belt of the Great Plains.

Xeric—Also called a Mediterranean climate with dry summers and cool, moist winters, soil moisture regime of California especially in the areas prone to winter landslides due to saturated soil conditions.

Classes of soil temperature regimes include the following:

Cryic—Very cold soils. Within this regime, the coldest soils have permafrost, temperature regime common to northern climates especially at higher altitudes.

Frigid—Cold winters, but summers are warm enough for crop production. Northern United States is an example.

Mesic—Warmer than frigid. In the United States, the Ohio River Valley is an example.

Thermic—Warmer than mesic. In the United States, the southern states are an example.

Hyperthermic—Warmest of the temperate zone soils. Found in the hottest parts of the continental United States.

Isohyperthermic—Hot tropical climate throughout the year.
(The prefix *iso* can be used with most temperature regimes if the soil temperature is quite uniform throughout the year.)

Soil Surveys

When the various soils on the surface of the earth are delineated on maps, it is called a soil survey. Usually the mapping of soils is done on a county basis.

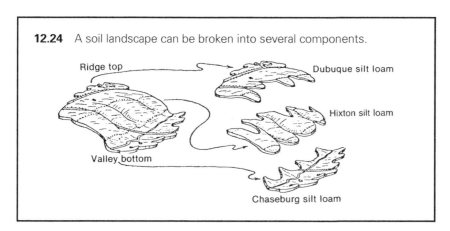

12.24 A soil landscape can be broken into several components.

Ridge top

Dubuque silt loam

Hixton silt loam

Valley bottom

Chaseburg silt loam

In the United States, the Natural Resources Conservation Service (NRCS) has the overall responsibility for making soil surveys to provide an inventory of our nation's soil resources. Agencies such as the U.S. Forest Service and the Bureau of Land Management also make soil surveys on lands for which they are specifically responsible. Through cooperative interagency efforts, the work by these agencies should blend in with that of the NRCS.

The soil scientists who make the soil maps are men and women who have graduated from a soil science program at an accredited university and completed a training period with an experienced soil surveyor. A knowledge of soil science for mapping purposes includes an understanding of geomorphology so that natural landforms can be identified. The boundaries of soil-mapping units commonly coincide with the boundaries of various segments of a landscape, such as ridges, side slopes, terraces, and floodplains. With sufficient experience in a locale, a soil surveyor should be able to predict the type of soil on any portion of that landscape and take soil cores just frequently enough to determine if the prediction was correct.

Figure 12.24 shows a small segment of a landscape (left) that has been expanded (right) to show its three component soil bodies. The Dubuque soil body has a silt loam surface horizon with clay subsoil over limestone bedrock. The Hixton soil body is a loam over yellowish-brown sandstone and siltstone. The Chaseburg is a deep silt loam formed in local alluvium.

A farmer becomes familiar with the soil components of the landscape on his farm after years of tilling the soil and digging holes in it. The sequence of soil bodies down a hillside is called a soil catena. In the example shown, all soils are well drained, although the water table is closer to the surface the farther a soil body is downslope. In less-hilly terrain in humid regions, a typical catena may consist of dry soils at the ridgetop and wet soils at the footslope position.

Landscape Patterns

The landscape is a mosaic of soil bodies that fit together as neatly as pieces in a jigsaw puzzle. Some landscapes have concentric, circular soil patterns; some are characterized by looped patterns; and others have parallel, linear soil patterns. The sketches in Figure 12.25 are simplified models based on a soil map, each representing 0.4 sq mile (1 km^2) of land in southeastern Wisconsin.

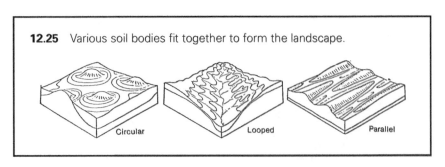

12.25   Various soil bodies fit together to form the landscape.

Circular   Looped   Parallel

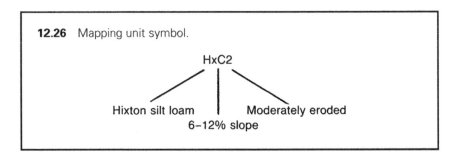

12.26   Mapping unit symbol.

HxC2

Hixton silt loam   Moderately eroded
6–12% slope

The patterns of landscapes influence farming, forestry, and wildlife management practices. For example, contour farming is not as easy on a landscape with tight circular patterns as it is in areas with linear soil patterns.

Making Soil Surveys

Since the 1930s, soil surveyors have been utilizing aerial photographs as a base map upon which boundaries between soil mapping units are drawn. The mapping units are determined by systematically traversing the land and augering samples of soil, each of which is checked by sight and feel for its distinguishing physical properties. Simple field tests may be conducted for pH, free lime, and soluble salts. The depth to which the soil is investigated will vary with its complexity of horizons, but it may be as deep as 80 in. (2 m). Each mapping unit in the pattern of soils is given a symbol for the unit it represents (Fig. 12.26). The soil surveyor interprets all of this information, delineates the soil type on a base map (Fig. 12.27), and moves to another location.

Soil mapping is as much a matter of photointerpretation as it is soil investigation. The tone of black-and-white photos gives a strong clue to the land form, degree of erosion of cultivated fields, the vigor of crops, areas of poor drainage, sometimes the species of natural vegetation, and much more to the trained eye.

Photo coverage for soil survey is made so that it provides stereoscopic coverage. This is accomplished by having a 60% overlap along the line of flight. When the photos are placed side by side, the soil surveyor can see photographed images for one position of the airplane with the left eye and another position with the right eye. This creates in the brain the appearance of a third dimension wherein objects with a higher elevation appear to rise up from the flat surface of the photos. In this way, the soil surveyor can properly record topographic features of the land.

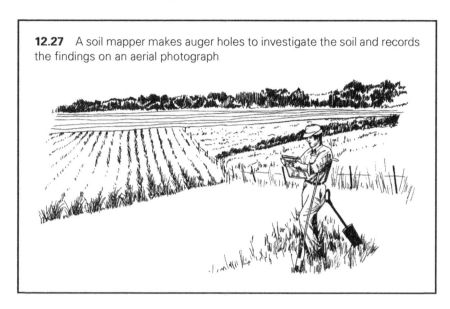

12.27   A soil mapper makes auger holes to investigate the soil and records the findings on an aerial photograph

Color photography has replaced much of the black-and-white film in recent years. In some places, color infrared photos of the land are available. They provide a better view of the pattern of vegetation and, by extension, the soil that favors a particular vegetative type.

The soil surveyor takes advantage of available geologic information. This is especially important when mapping soils where the parent material is residuum or bedrock as opposed to that which was transported by wind, water, or glaciers. In residuum, bedrock strongly influences the type of soil that will develop.

The application of computers has provided soil surveyors with improved base maps by removing distortions inherent in photos made from an airplane. These corrected photos are called ortho (true) photos. When laid side by side they match. They can be digitized so that the photo image can be reproduced on a computer screen and zoomed in or out for a detailed or general view.

The global positioning system (GPS) was introduced in Chapter 8. It is possible for soil surveyors to carry a GPS locator and a compact computer loaded with a digitized three-dimensional base map. In this way, more accurate information will be entered into a database as the fieldwork progresses. Experience has shown that such technical advances improve the quality and efficiency of soil surveying.

Uses of Soil Surveys

In the United States, the NRCS, agricultural universities, and county extension offices can usually supply information about the soils of a given county. For example, for Columbia County, Wisconsin, there is a 156-page published soil survey report with 122 map sheets showing soil mapping units in the county. Each unit has its own combination of soil horizons, slope, and moisture regime characteristics. As an example of the kind of information gathered, two of these soils are described in Table 12.4.

Just as the making of soil maps is changing, so is the publication of soil survey reports. Recent hard copy reports are more flexible, with only the photo section bound and the interpretive section in loose-leaf form for ease of modification. There is getting to be less

Table 12.4 Two soils of Columbia County, Wisconsin

| Soil name | Kind of horizon | | | | Typical corn yield in bushels per acre[a] |
	Topsoil (A)	Subsoil (B)	Parent material (C)	Slope	
Plainfield loamy fine sand	Thick, pale (ochric)	None	Sand, acid to neutral	Undulating (2% to 6% gradients)	45[b]
Plano silt loam	Thick, dark (mollic)	Sticky to firm, blocky (argillic)	Silt over sandy loam glacial deposit	Nearly level (0% to 2% gradients)	130[c]

Without irrigation.
45 bushels per acre = 2,800 kg/ha.
130 bushels per acre = 8,160 kg/ha.

emphasis on hard copy in the form of books containing photos, printed discussion, and interpretive tables. More emphasis is being placed on digitized photos and computerized interpretive information. This will extend the useful life of the reports because they will be easily updated and tailored to the specific needs of future users.

In the United States, soil surveys have been published for most of the agricultural regions as well as much of the forest and rangelands. In some cases, counties surveyed more than 30 years ago have to be rechecked to gather information that was not collected at the time of the original survey. The revised surveys are being digitized in accordance with the Soil Survey Geographic (SSURGO) database and must meet the national carto-graphic standards. As these efforts progress, more soils information will be added to what is currently available on compact discs and Web sites.

Examples from the map section of a traditional soil survey report of Randall County, Texas, are shown here to illustrate some of the information available. Figure 12.28 shows a portion of a detailed soil survey map that might be used as the basis for making a farm plan. Figure 12.29 is a generalized map of the entire county suitable for making decisions about broad areas of crop and range management.

Soil surveys no longer benefit only agriculture but are of value to anyone who makes decisions about the land. This includes farmers and ranchers who want to maximize pro-duction efficiency. Fertilizer dealers better serve their customers through an understand-ing of the soils being managed. Engineers who bid on earth-modifying projects such as roads and airports find soil surveys useful for planning. Land developers must consider the soil for foundations, streets, lawns, and sometimes the septic systems. Bankers and other money-lending agencies can get a better feeling for the security of their loans if they know the potential of the land being used by the borrowers. Foresters use soil maps in species selection for regeneration, and they also plan the harvesting operations based, in part, on the bearing capacity of the soil and its susceptibility to erosion. Parks and other recreational facilities must be planned around the soil's suitability to support human and vehicular traf-fic. Tax assessors and appraisers are becoming increasingly aware that they can make more equitable assessments on farmland and ranch land if they take advantage of the information provided in soil survey reports. As can be seen, there is a wealth of information available in these reports; only a few have been given here to illustrate the point.

Even if there is a published soil survey with a complete map of an area, it is a good idea to examine the soils. There is more variation on the landscape than the published map can show. It is helpful to learn how to recognize the properties of the specific soil horizons.

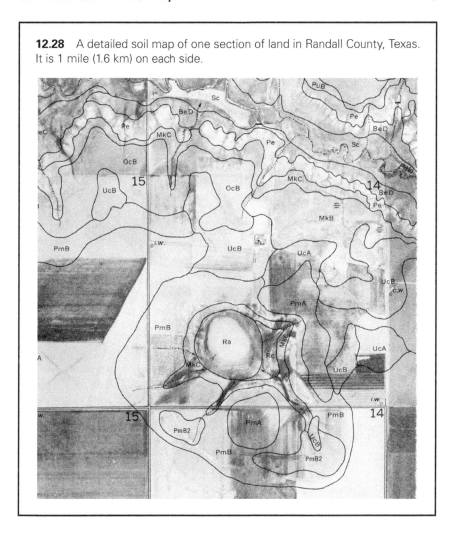

12.28 A detailed soil map of one section of land in Randall County, Texas. It is 1 mile (1.6 km) on each side.

Land Capability Classes

Soil survey maps are interpreted for many uses in the soil survey report, but in agriculture they are used most extensively to determine land capability classes. All soil bodies are placed in one of eight classes. Class 1 land is easily managed for crop production without having to overcome any appreciable limitations. Classes 2 through 4 have increasing limitations if they are to be tilled for crop production.

Land in classes 5 through 8 is not recommended as cropland, and in class 8 the primary value is aesthetic and as a watershed.

The four possible subclasses—E: erosion, W: wetness, S: rooting zone (e.g., shallow), and C: climate (e.g., arid)—show the dominant limitation that causes the soil to be placed in a particular class. The most obvious to cropping is the erosion hazard, which is based on slope.

12.29 A generalized map of Randall County, Texas.

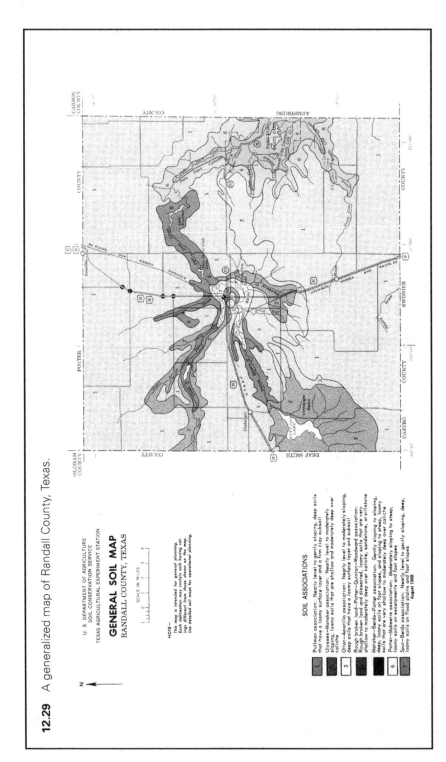

Several drawings in other chapters can be used to illustrate some of the land capability classes. Class 1 land is depicted in Figure 3.8 because the land is level and no crop-limiting characteristics are shown. Figure 6.15 might be classed 2W because of its need for drainage. The erosion problem in Figure 10.2 is present because class 3E land is being improperly tilled up and down the hill. The shallow soil over bedrock in Figure 10.1B could be in class 6S, and the bog in Figure 12.9 might best be in class 7W.

Land has to be evaluated for several possible limitations to place it in the proper capability class, but by looking at these examples the reader should be able to understand better how the system works. Similar systems are used in the soil survey report for the land's potential for forestry and wildlife habitat.

Soil Landscape Appreciation

The eminent biologist Aldo Leopold wrote in his book, *A Sand County Almanac,* "When we see land as a community to which we belong, we may begin to use it with love and respect." It helps to know enough about the soil to evaluate soil landscapes intelligently and avoid mistakes in land management decisions. Landscape appreciation for its own sake is of interest to the ordinary citizen who does not have direct responsibility for soil management but feels an affinity to it.

Soil Landscape Appreciation

CHAPTER **13**

Soil and Its Uses

Soil is a medium in which plants are grown for food and fiber—the principal use of soil. It gives mechanical support for plants roots—a necessary function. Soil acts as a decomposer for vegetative and animal remains—a necessity to keep organic materials from continuing to accumulate on the surface. Soil is involved in water and air movement—vital to plant growth. In addition, soil is beautiful. It is an aesthetic resource which is pleasing to many people.

In addition to these general functions of soil, there are many more specific uses of soil other than for producing food and fiber. They range from foundations for structures (roads and buildings) to use in water treatment and waste disposal facilities by municipalities to its use in urban landscapes for lawns, flowers, and vegetable gardens to dams, levees and ponds.

The purpose of this chapter is to discuss the various uses of soils when used for purposes other than for agricultural production. Chapters 3 through 7 provide the properties and characteristics of soil that can be used in its management whether in crop production, in urban landscapes, or for foundations for structures or its many other uses.

Principles of soil management in Chapter 9 are mainly for agricultural production, but they can also be applied to many other uses of soil.

Urban Soils

As population increases there are not only many more people to feed but also many more people to house. As a result, land once in food production is being taken by expanding cities and suburban development. As urbanites populate the communities that include their homes, work places, industry, and infrastructure, the natural soil body will begin to undergo changes to suit each owners needs with only some regard to the consequences of such changes to the soil or the broader environment.

Soils are soils: they are a natural body of matter that once were left only to the forces of nature to evolve into what they are today. An urban soil is one that has been disturbed, modified, or contaminated through human activity and urban/suburban development. They no longer have characteristics like the natural landscape surrounding the urban area.

Soil Science Simplified, Sixth Edition. Neal S. Eash, Thomas J. Sauer, Deb O'Dell, and Evah Odoi.
© 2016 John Wiley & Sons, Inc. Published 2016 by John Wiley & Sons, Inc.

The modifications to the urban soil have brought about significant **variability** in soil developmental characteristics due to changes in virtually all of the soil-forming factors compared to natural soils. The **hydrology** of an urban area also changes, resulting in an effect on plant growth. The intensity of these changes will depend on past land use and or site-specific human activity.

Variability and Hydrology

The one common theme of urban soils is the degree of lateral and vertical **variability** over short distances. While lateral and vertical variability is common in natural soil profiles, as the land begins to become more urbanized there is a need to excavate soil to install underground utilities, foundations, basements, roads, or change elevations to accommodate design features of various structures. As a result soils that are excavated may have material removed and placed at a different location on-site, hauled off site, eroded, back-filled, or hauled in from another location. In addition, human artifacts used in construction are commonly found buried in urban soils and can be a problem when planting shrubs, trees, and so on.

In some cases the natural soil is buried under on-site or off-site materials, while in others it is mixed with soil from a different location or depth. Urban soils are disturbed soils, made up of a mixture of soil materials obtained from one or more locations and placed in layers of varying thicknesses making it difficult to predict their horizon characteristics. The extent of disturbance will depend on the existing soil properties, the intended use and reason for disturbance, and the properties of external soil materials hauled in.

There are many areas of urban land that remain relatively undisturbed and retain their natural soil characteristics. These undisturbed areas are typically represented by urban forests, or areas beyond the influence of the construction zone. These various effects create a "mosaic" of soil conditions, ranging from natural to highly disturbed "anthropogenic" soil profiles.

The hydrology of an area can change considerably under the impact of urban development. This in turn impacts plant growth. Two hydrological aspects that are important are **infiltration** and **overland flow.**

Urban areas are characterized by impervious surfaces (roofs, driveways, parking lots), layered subsoils, modified slopes, and land surfaces with or without vegetative cover. Rainwater falling on impervious surfaces may or may not be distributed evenly on the adjacent land surface; instead it may be collected from roof tops and sump pumps and discharged in localized areas on the land surface or may be directly discharged to a storm/sanitary sewer. When the water collected from roof tops and other impervious areas is allowed to run freely on to the adjacent land, water will soak into the soil in an uneven manner (altering the subsurface water movement). Excess water that does not infiltrate the soil is distributed farther down slope from the point of discharge. Water available for infiltration immediately adjacent to the discharge points of the paved areas depends on the slope characteristics and the clay content of the area. In soils where infiltration is rapid, the amount of water entering a soil profile is higher than from a natural rain event. Compared to rural soils, localized urban soils are exposed to more infiltration water than that which would have resulted from natural precipitation in the area.

When the additional water percolates through the soil, it not only impacts water availability to urban vegetation but also impacts future horizon development. Additionally, the excess water can help move soluble chemicals farther down the soil profile and eventually into the groundwater. During short rain events, rainwater usually soaks into the soil plus any overland flow that comes from paved surfaces. During intense storms, the infiltration rate of the soil may not be rapid enough to absorb all the rain; therefore, more overland flow is created from unpaved surfaces.

Restoring soil structure by the addition of organic amendments such as compost and mulches are known to increase infiltration rates by a factor of 6-10 times than that of a similar surface soil without the amendments.

The hydrology of an area can also have an impact on **flooding.** Modifications can be made to lessen the impact of excess water. In most urban soils, the surface slopes (aspect and steepness) are modified to efficiently remove excess water along the land surface. Slope modifications can change the quantity and the direction of water flow. Often drainage ditches, swales, roof drains, curb, and gutters are designed to carry water rapidly away from areas of human activity immediately after a rain event. Additionally, the rapid removal of water from locations in the upper elevations of a watershed would mean potentially excess water in lower elevations. Adequate precautions must be taken to avoid flooding of low-lying areas.

Many urban planners will discourage property development in urban flood zones because of a periodic serious risk of being flooded. Land in flood zones is normally near water, often has flat slopes and therefore preferred for construction of homes and offices. Many urbanites have chosen to build on such sites and installed expensive flood control measures. With the hydrology of the higher elevations of the watershed constantly changing due to development, the flood control structures often turn out to be ineffective subjecting the property owner to periodic flooding and loss of life and property. The proximity to surface water on flood plain developments also has a negative impact on surface water quality.

Soil Properties

In addition to the lateral and vertical variation in materials that remain in place in an urban soil environment, the physical, chemical, and biological properties of the soil are significantly altered when soils are disturbed. All these have an impact on urban soil quality and the environment.

Physical Properties. The physical properties of soil that are altered include texture, compaction, temperature, water storage and movement, air, structure and porosity, and slope. The first three listed are discussed in greater detail. Erosion, a function of many of the physical characteristics of soil, is also discussed.

Texture in urban soils will vary significantly in disturbed conditions. The texture of the material placed in the ground will depend on whether it is of natural soil origin or man made (such as ashes, demolition waste or dredged material from a pond or wetland). Sometimes the choice of these materials is based on cost and a specific soil property intended to be altered (drainage, elevation changes, plant growth, etc.). In specific locations where the local soil does not meet a specific engineering need such as physical support (road bed, driveway), water movement or retention may be desired (gravel, sand, or clay may be hauled in). For intensively landscaped areas, soil and amendments rich in organic matter may be hauled in thus modifying soil texture in the surface or subsurface layers.

Compaction in urban soils varies significantly and depends, for example, on the use of heavy equipment and intensive pedestrian or vehicular traffic during and after construction. Compaction is intensified when these activities are carried out on wet soils. Compacted soils will affect root penetration, air movement, water movement, and the amount of stored water available for plant or microbial growth. Such impacts are obvious in areas covered by lawns that are consistently subject to human foot traffic or vehicular traffic. Soil paths get so compacted that the roots have difficulty growing in the soil, and water has difficulty infiltrating the soil causing puddles after a rain event.

Compaction can occur in excavated or unexcavated soils from the weight of construction equipment. Often the soils may be compacted deeper into the profile affecting the structure, porosity and density of the soil. While surface compaction can be undone by tilling to a certain depth, it is more difficult to undo the compaction at deeper soil depths. Such deep compaction can affect urban tree growth, often stunting growth (compared to their counterparts growing in undisturbed forested conditions). One of the causes for stunted growth could be a lack of oxygen or physical restrictions (smaller volume of pores) for root growth.

When soils are compacted deeper in the profile, water penetration can be affected resulting in a build up of a perched water table affecting tree root penetration and oxygen needs of tree roots. Such a soil situation could result in a shallow root system and potential for tipping over under strong wind conditions. In other urban situations, the land adjacent to an existing tree trunk may be covered with additional soil burying the feeder roots deeper and cutting off oxygen supply to the roots.

It may take several years before the subsoil in an urban area returns to its natural density. The uncompaction process will depend on the texture of the compacted soil material, the rooting habits of the vegetation, the frequency of wetting/drying or freeze/thaw cycles, and the extent of additional compacting activities at the surface. Compaction, however, is not of concern in well-managed flower beds and turf grass areas where there is little human traffic after the ornamental plants have been in place. These flower beds may also periodically receive organic matter additions through mulch or compost.

Soil temperature will be impacted by urban development. Concrete surfaces will absorb solar radiation and retain heat that may be later released to the surrounding soils. In addition heat released from human activities tends to make urban environment and urban soils warmer than soils in nonurban areas, thereby, extending the growth cycle. Urban area soils may see fewer frost-free days extending the growing cycle and increasing the decomposition rate of organic matter thereby affecting nutrient flux. Tall structures may have a shading effect on one or more sides of the structure making the soil cooler in areas protected from the sun. Urban effects of shading from buildings is likely to modify moisture and temperature regimes in isolated situations. Application of mulches in landscaping will also have a similar cooling effect impacting organic matter decomposition and nutrient release in the shaded areas.

Erosion in urban areas can be a problem. There is a significant potential for erosion when a bare soil is exposed to wind and rain, especially if the subsoil (with little or no organic matter as binding agents) is exposed to the elements. Erosion at constructions sites could be a serious problem amounting to 100 times more erosion than when the bare A horizon of the same soil is exposed to the elements. Erosion in urban areas is further accelerated due to alteration of slopes when the soil is excavated and placed in a pile with steep side slopes. Precautions should be taken to prevent the soil being washed away and deposited on neighboring property. Often silt fences are placed at construction sites to trap the soil yet allow the water to pass through the fence and prevent any deposits on adjacent land. Excavated bare soil exposed to prolonged dry periods can result in serious dust issues on windy days. It is often recommended that earth-moving activities be coordinated with appropriate weather conditions. If bare soil must be left exposed for prolonged periods of time, all precautions must be taken to somehow bind the soil by hydromulching, spraying water on dry surfaces, or establish a vegetative layer at the surface of the disturbed soil.

Once an urban soil is landscaped, the slopes meticulously modified, and the land surface is covered with vegetation during most times of the year, these soils tend to be less erosive than their natural counterparts under agricultural land use.

Chemical Properties. There is a higher degree of chemical variability in urban soils as compared to rural soils. Two major reasons for such variability are the disturbed/mixed nature of urban soils and the extent of chemical addition from external sources.

Among the chemical properties altered are pH; ionic concentrations of the soil solution; nutrient balance and fertilizer additions; organic matter content; addition of air pollutants from cars, industries and the burning of fossil fuels; and possibly contaminants released to the soil from human activity and manufacturing industries. Adding to the alteration of chemical properties is foreign matter such as concrete or construction materials, ashes, pesticides, soil amendments, and the disposal of human wastes via septic systems.

Urban environments receive higher concentration of atmospheric chemicals in the form of dry or wet atmospheric deposition. The sources of these chemicals are burning of fossil fuels, emissions from industry, and exhaust from automobiles. Atmospheric temperature inversions in urban environments often keep these chemicals suspended over the urban areas. Heavy metals such as lead, copper, nickel, and air pollutants such as sulfur and nitrogen compounds are often present in these deposits. These chemicals accumulate in the surface horizons of urban soils.

The **pH** of urban soils is subject to many factors. Many of the air pollutants may lower the soil pH in soils that have a low buffering capacity. The acidic conditions promote leaching of bases in urban soils. Heavy metals present in urban soils could be released at lower pH and absorbed by the plant. In extreme situations some of the heavy metals can reach toxic concentrations for plants and animals. Soil pH can vary between localized areas managed for turf grass such as transportation and commercial corridors and those around residences. Soils in contact with or close proximity of building and paving materials rich in calcium tend to exhibit a higher pH. Often lime additions are practiced in humid regions to maintain an adequate soil pH. The higher pH limits heavy metals from coming into solution; therefore, they remain locked into the soil matrix and reduce the risk to plants and animals.

The levels of **plant nutrients** in urban surface soils are generally found to be adequate because in most instances a certain amount of topsoil is placed on disturbed land. In intensively managed landscapes where fertilizer, compost, and organic mulches are added on a regular basis, nutrient levels may be at higher than normal levels compared to rural soils. In some urban soils repeated fertilizer additions containing a disproportionate amount of one nutrient can often create an imbalance in nutrient availability to plants. The overapplication of even a balanced nutrient source, while helping vegetation to grow faster, may have a negative impact on surface and ground water. Excess phosphorus in urban lakes and streams has been traced to excessive fertilization of lawns with phosphorus. This impacts biodiversity in the receiving water. Evidence shows that overfertilization with nitrogen combined with irrigation (practiced often in urban environments) may lead to flushing of nitrogen into the groundwater.

Addition of **soluble salts** is also a factor in urban soils. Irrigation water, excessive use of inorganic fertilizers, salts used for deicing paved surfaces (streets, driveways, and sidewalks) in cold climates, water softeners, and so on, have an impact on the soluble salt content of the soil water and the vegetation growing at that location. In areas where the salt concentration is high, salt tolerant plants gain a competitive edge. Salt concentration will also impact the types of microorganisms that will be active in the soil.

Heavy metals in urban soils come from various sources. Atmospheric additions, original rock (copper, mercury, lead, zinc), and past industrial/mining activities (cadmium, nickel, copper, lead) serve as potential sources of heavy metals in the soil. Lead in particular is common in urban soils as a result of automobile exhaust when lead was part of gasoline, and lead paint used in the exterior and interior of old buildings (prior to 1978). Urban fires and dredged marine sediments also add metal contaminants to local soils. Presence of metals does not preclude plant growth, only in situations where toxic levels accumulate will there be no vegetative growth.

Extraneous chemicals can be a problem at times for urban soils. Modern industrial societies have developed new organic and inorganic compounds for various industrial, agricultural, and manufacturing applications. Some of these chemicals are inert and harmless while many of these can be extremely toxic to humans and other organisms. Some of these chemicals are not biodegradable. Many have accidentally made their way into our environment (soil, water, air, and plant and animal life). The intensity of human activity and use of these chemicals by humans in various applications leads to increased concentrations of such chemicals in urban soil environments than in rural environments.

The soil is often expected to miraculously decompose, alter, treat, detoxify some or all of these chemicals and somehow make the soil environment unharmed and yet have little impact on surface and groundwater. Once these materials enter the soil, however, they affect all forms of organisms in one way or the other and impact the overall ecological cycle in these environments. Each soil organism has a certain threshold concentration below which its activities are minimally affected. Once these limits are exceeded, ecological damage is enhanced.

Biological Properties. The **organic matter** content of urban soils is highly variable and will depend on several factors including intensity of landscape management; additions of organic amendments; degree of soil disturbance; extent of organic matter contributions from vegetation; and rate of organic matter decomposition. Litter and organic matter decomposition rate is dependent on the soil environmental conditions, microbial and invertebrate population present, and the source/type of organic matter. Typically, organic matter decomposition rates are faster in urban soils than in rural soils. This is attributed to the higher temperatures and adequate amounts of nitrogen contributed from fertilizer additions as well as atmospheric deposition. Other studies have shown that due to the higher ozone concentration in urban air, the vegetative matter exposed to ozone is subject to slower decomposition rates.

Soil organisms in urban soils are impacted by soil conditions. The consequence of changes in soil physical and chemical properties in urban areas means that the habitat for native species may have changed allowing non-native species of plants and animals to become dominant. The fragmented nature of urban landscapes reduces the habitat available for any given animal and plant species allowing native species to be displaced and replaced by nonnative species. Many non-native species are planted in urban areas that may perform well under disturbed soil conditions while the native species may not be able to do as well under the disturbed soil conditions.

Sometimes these plant species are introduced via seeds present in soil or compost material hauled in from external sources. At other times plant species are part of a landscaping plan. Unless the plant can handle the physical and chemical conditions of the soil, it may not thrive. The interaction of soil, plants, insects, and microorganisms among each other can determine the soil nutrient flux under the new environment.

Growing Plants in Urban Landscapes

To obtain optimum growth of plants in the urban landscape, proper management practices are just as important as with the production of food and fiber. The ideal situation is to establish and maintain the correct combination of all soil factors necessary to optimize the growth of plants.

The **physical condition** of the soil is the starting point. Starting with and keeping the soil in good physical condition means that it will be easy to till and that seed germination will be optimum or that transplants will resume growth quickly. The rate at which water soaks into the soil and the water-holding capacity of the soil will be optimal. Root growth and soil-air interchange will benefit from good physical condition of the soil. Applying and incorporating organic materials is one way to assure good physical condition.

Next the **chemical characteristics** of the soil need to be in proper balance, which means maintaining soil fertility levels and the correct pH. Start with soil pH—the acidity or alkalinity of the soil. Many plants can achieve optimum growth over a wide range of soil pH. Some plant species, however, prefer an acid soil condition while some will perform best in an alkaline soil condition. Apply limestone to correct the pH of soil if it is too low, or use sulfur for a soil that is too alkaline. Limestone and sulfur needs can be determined by soil tests.

Next assure a proper balance and supply of plant nutrients. If needed, apply fertilizer. It can provide the plant nutrients necessary for good growth provided it includes those needed to balance the nutrient levels in the soil and to provide the nutrients needed by plants. Fertilizer may contain only one nutrient or any combination of the 14 elements required for growth (other than C, H, and O).

Nutrients other than nitrogen normally need to be applied prior to planting or transplanting because they normally move very little in the soil after application. An exception is potassium in sandy soils. High nitrogen-use plants, such as grasses, may need a relatively large amount applied at seeding or sodding while for low nitrogen-use plants, such as flowers and shrubs, nitrogen can be applied minimally at planting and top dressed as the season progresses. Apply plant nutrients according to soil tests.

Soil amendments such as limestone and organic materials should be applied prior to planting. It would be convenient to apply these at the same time that fertilizer is applied. Amendments and fertilizer should be incorporated into the soil before planting or transplanting.

The **biological characteristics** of the soil are important for good growth. Organic matter and humus need to be maintained at an adequate level so that maximum benefits result. Materials that serve this purpose are peat moss, compost, manure (preferably treated and partially decomposed), or almost any type of plant residue (leaves, grass clippings, etc.).Organic materials need to be incorporated prior to seeding, sodding, or transplanting.

As with any crop in agricultural production, proper selection of a variety or species is desirable, whether it is trees, grass, shrubs, or flowers. Water needs should be considered. For example, in the low-rainfall area of the Western United States, use species of grass that have a low water requirement such as Buffalo grass. In the case of trees, shrubs, and ornamentals, a Xeriscape approach is often desirable.

Keep in mind that the needs of soils used in urban landscapes are the same as those in agricultural production. Chapter 9 on Soil Management gives a much more complete discussion on soil management, most of which applies to turf, trees, ornamentals, vegetable gardens, and shrubs.

There are many aspects of landscape soils and plants and their management that often require professional assistance in soil preparation and in planting. This is particularly true for turf planted as sod and for trees and shrubs. For a long-term use situation, it is often better to ask for professional assistance to assure long-term success.

Predicting soil behavior of an urban soil is difficult because the physical, chemical, and biological changes brought about by one landowner to suit their own needs may not serve the needs of a new landowner. These changes in land use or modification of soil conditions are not always accurately recorded and transmitted to the next landowner. Therefore, it is critical that a prospective landowner should exercise due diligence in evaluating soil conditions that meet the needs of a specific project.

Engineering Uses

Soil is a source of material that has a wide array of uses for engineering purposes. Some examples include fill for dams and levees; foundation material for roads, runways, and buildings; aggregate (sand and gravel) for making concrete; clay for sealing the bottoms and sides of ponds, canals, and solid waste landfills; cover material over tanks, utility lines, tunnels, culverts, and conduits (sewers, drains, pipes for water, oil, and gas); and a porous medium for treating liquid wastes.

Soil has been used to make homes like the adobes of the Southwest and the sod houses of pioneers settling on the American prairies. Soils support enormous loads, both inanimate (roads and buildings) and animate (people, animals, and plants).

To the engineer, soil is any surficial material of the earth that is unconsolidated enough to be dug with a spade. Soil mechanics is the field of engineering devoted to the use of soil as a building material. In the vocabulary of engineers, soil includes both the soil of the soil scientist and any loose substratum that may be present. The concepts of the soil as considered by the scientist and the engineer are merged in this section.

An advantage of using soil for engineering purposes is that there is so much of it, and it may be already on the site, which avoids the expense of hauling in other material. Another advantage is that soil can be so readily shaped into almost any desired form. Depending on how it is manipulated, soil can allow the passage of water through it or it can be made almost impermeable.

There are also some disadvantages for using soil in engineering. Soil is extremely variable, both geographically and over time. Cycles of wetting and drying as well as freezing and thawing change the engineering properties of soil. Unlike known types of steel or wood, soil is not a uniform material for which reliable strengths can be computed. Stable dry loam may be adjacent to unstable wet clay in a lowland. During a rainy period, the dry loam may also become wet and unstable. In winter both soils may freeze and heave in such a way as to crack pavements and basement walls, especially where the moisture content is high. Table 13.1 compares the suitability of three soils for various engineering uses. Even though these soils are usually found next to each other in the landscape, their suitability for different engineering applications varies widely.

Engineering Properties of Soils

Preceding chapters have discussed soil physical properties from the perspective of factors relating to crop production. Soil mechanics also deals with physical properties as they relate to the use of soil as a building material. Some physical properties like particle size distribution and bulk density are important to both soil scientists and engineers. Even so, engineers have developed different soil classification systems specifically for engineering applications (Fig. 13.1).

Table 13.1 Suitability or limitation rating for soils of the Clarion-Nicollet-Webster association

	Clarion	Nicollet	Webster
USDA classification	Fine-loamy, mixed, mesic	Fine-loamy, mixed, mesic	Fine-loamy, mixed, mesic
	Typic Hapludolls	Aquic Hapludolls	Typic Haplaquolls
Shrink-swell potential	Low Good	Moderate	Moderate
Roadfill		Fair (wetness and low strength)	Fair (low strength, wetness, and shrink-swell)
Embankments, dikes, and levees	Severe (piping)	Moderate (piping)	Severe (wetness)
Dwellings with basements	Slight	Moderate (wetness)	Severe (wetness)
Septic tank absorption field	Slight	Severe (wetness)	Severe (wetness)
Sewage lagoon area	Moderate (slope and seepage)	Severe (wetness)	Severe (wetness)
Sanitary landfill area	Slight	Severe (wetness)	Severe (wetness)
Daily cover for landfill	Good	Fair (wetness)	Poor (wetness)

Source: Nelson, G. D. 1990. *Soil Survey of Murray County, Minnesota.* USDA-Soil Conservation Service, U.S. Government Printing Office, Washington, DC. Reasons for limitations are given in parentheses

13.1 Soil classification systems used by engineers (AASHTO and USC) have different ranges for particle size distributions than the USDA system.

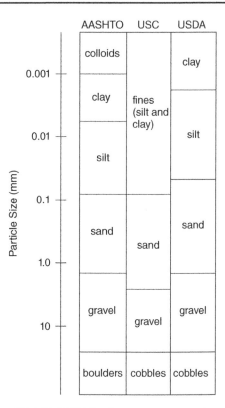

The Unified Soil Classification (USC) System was developed during World War II for the construction of military airfields and has subsequently been modified for use in foundation engineering. The American Association of State Highway and Transportation Officials (AASHTO) System is widely used by state transportation departments and the Federal Highway Administration for the design and construction of transportation lines. Both the USC and AASHTO classification systems include several tests in addition to particle size distribution.

Two standardized tests are often completed to test a soil's suitability as a building material. These tests are called the Atterberg limits, or the liquid and plastic limits. At a high water content, a soil possesses the properties of a liquid. As it dries, it acts more like a plastic, then like a semisolid, and finally like a solid when it is dry. The liquid and plastic limit tests are completed to identify the moisture content at which a soil changes from the consistency of a liquid to a plastic (liquid limit) and from a plastic to a semisolid (plastic limit).

Table 13.2   Characteristics of soils for engineering purposes

Kinds of Information about soils	Comments
Soil texture	The inorganic (mineral) part of soils is a mixture of sand, silt, clay, and coarse fragments, even including boulders. the USDA, USC, and AASHTO systems are compared in Figure 13.1.
Kinds of clay	Clay species vary in degree of shrink-swell potential and other activity.
Depth to bedrock	Very shallow soils are usually unsuitable for excavation for basements, ditches along roads, or utility lines.
Kinds of surficial bedrock	Bedrock may be very hard (granite) or porous (sandstone, shale).
Soil density	Soil horizons range in density from porous to cemented. The denser soils have the higher bearing capacities and rate of transmission of vibrations (sound may travel twice as fast through dense soil as through air).
Content of rock fragments	Soil with many rock fragments is usually difficult to excavate and compact uniformly.
Erodibility	Many sandy soils are susceptible to wind erosion. Silty soil gully easily. Some clay soils are subject to piping, which is subsurface erosion by spontaneous tunneling.
Surface geology	The lay of the land affects land use. Proportions of steep and level land vary as well as soil pattern (linear, circular)
Soil pH (reaction)	Degree of acidity or alkalinity influences soil behavior physically, chemically, and biologically. To stabilize soil, engineers sometimes add hydrated lime to it, which raises the pH into the alkaline range.
Salinity	High salt content of soil affects its stability and that of vegetative cover.
Corrosivity	Soils differ in capacity to corrode buried pipes and tanks. Wet, acid soils are usually very corrosive
Depth to seasonal water table	Soils with a seasonally high water table provide inadequate support to roads and structures. Frost action is most severe in wet soils.
Plasticity	Clayey soils are commonly quite plastic and, with wetting, become fluid like a thick liquid. The "Plastic index" is the range of percent moisture content in which a soil is plastic.
Content of organic matter	To support growth of protective sod, a soil layer containing organic matter is needed. Otherwise, organic soils, paritcularly peats and mucks, are usually removed in engineering projects.

The Atterberg limits help engineers decide whether the soil material under consideration is suitable for their project or whether a different soil is needed. Liquid and plastic limit values are used with particle size information and other tests to classify soils in the USC and AASHTO systems.

Some of the soil characteristics that should be known before decisions about engineering uses are made are given in Table 13.2. Many soil survey reports include tables showing the suitability of the different soils for various engineering purposes. The American Society for Testing and Materials (ASTM) publishes official methods for measuring engineering properties of soils. It is important to understand the soils of an area by studying soil survey maps and reports, talking with soil scientists and engineers, and making the proper measurements.

Roads, Residences, and Structures

Nearly all roads and small buildings are placed on soils, many of which are soft in wet seasons. In some areas subject to seasonal frosts, the soil heaves (lifts) and dirt roads may become impassable in certain seasons. Modern engineers find that naturally well-drained, very sandy, and gravelly soils provide the most trouble-free bases for roads and building foundations. On clay soils, a blanket of sand and gravel is placed over the clay before pavement is laid, and ditches are dug on either side to drain away storm and seepage waters (Fig. 13.2).

Water, so essential for plant growth, is often undesirable in soils on which structures are placed. The long life of a road or building foundation depends on maintaining soil conditions that permit it to behave like a compact, well-drained sand or gravel.

Roadside soils also perform special functions unrelated to their engineering uses. Certain organisms in soil can metabolize various components of automobile exhaust. Embankments of soil along major highways absorb much of the sound of traffic, thereby reducing noise pollution in the area.

The high capacity of sand and gravel to support weight arises from the particle-to-particle contact without lubrication of silt and clay between them. Gravel made of strong and stable quartzite is long lasting under the stress of heavy traffic on pavement above the gravel bed. Concrete pavement itself contains as much as 50% gravel and 25% sand.

Dams and Levees

Soil is the most readily available and least expensive material for the building of dams and levees. Numerous reservoirs, lakes, and ponds back up behind dams largely made of

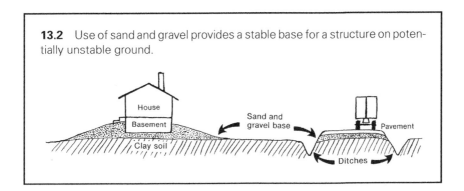

13.2 Use of sand and gravel provides a stable base for a structure on potentially unstable ground.

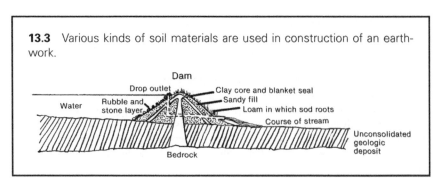

13.3 Various kinds of soil materials are used in construction of an earthwork.

soil. Many cities located in floodplains are defended against overwhelming floods by miles (kilometers) of levees built of soil.

Such earthworks need to have two qualities: stability and impermeability. Four kinds of soil materials are needed to achieve these properties in a dam or levee: (1) a clay core and blanket are compacted and kept moist as a seal against leakage of water; (2) a sandy mass is added, surrounding the clay, to drain water away; (3) a layer of stone and rubble (riprap) is piled on the surface exposed to moving water such as waves and river currents; and (4) a good loam is used to cover remaining surfaces of the earthwork to support the growth of protective vegetation (Fig. 13.3).

Sand is the most easily excavated and transported construction material because it is loose and does not become sticky upon wetting or hard upon drying. Bagged sand is used in emergency enlargement of levees during exceptional floods. Because reservoirs and river floodplains tend to gradually fill with sediment washed in from upstream, this material often must be removed from the reservoir or channel (by dredging) every 20–50 years or so. The dredged sediment can be used to build artificial islands or protected mounds on the floodplains. It can be returned to the farmland but usually at great expense. Reducing erosion is therefore doubly important for both soil conservation and reducing sediment buildup in water bodies.

Ponds and Canals

Bottoms and sides of ponds and canals commonly need to be sealed to prevent leakage. Various kinds of lining materials may be used that are less expensive and less bulky than concrete. If clay is available locally, it may be mixed with bentonite, a special type of swelling clay, to make a tight clay liner. Fabric liners containing plastic or rubber or coated with asphalt may be used in the absence of clay.

Underground Structures and Lines

Earth sheltering of homes and burial of utility lines (Fig. 13.4) and tunnels protect structures and facilities from unfavorable temperatures that take place at or near the surface of the ground. However, some precautions should be followed.

Underground installations must be designed to support the great weight of soil cover. The strain on buried structures from the weight of overburden may be heightened in wet seasons by expansion of clays in the soil. Seepage of water into cracks in buried structures may be a recurrent problem. Growth of deposits of calcite or iron oxides in cracks may gradually shatter concrete below ground.

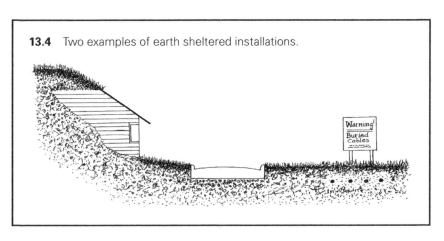

13.4 Two examples of earth sheltered installations.

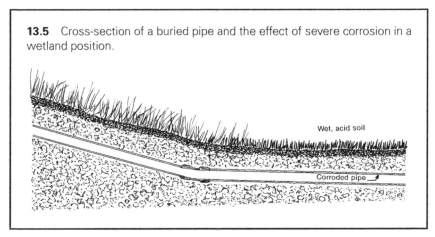

13.5 Cross-section of a buried pipe and the effect of severe corrosion in a wetland position.

Concrete and metal pipes and tanks are subject to corrosion in some moist soils (Fig. 13.5). Tiny electric circuits may develop spontaneously between the soil and iron pipes in such a manner as to literally bore minute holes in the pipes by dissolving the metal. In regions of permafrost, water mains must be insulated against freezing both from above and below.

Municipal Waste

Waste Treatment

Human activities generate liquid and solid wastes that require treatment and/or containment. Liquid wastes include residential and industrial wastewaters. Solid wastes include household and office wastes that consist of yard waste, paper, plastic, glass, and metal, and industrial wastes that include manufacturing by-products such as sludges from paper mills. Many types of liquid and solid wastes can be disposed of in soil.

Soil is a porous medium with enormous internal surface area populated by microorganisms that are capable of decomposing biodegradable materials. Because of this characteristic with microorganisms, when biodegradable wastes are disposed of in soil, they are broken down and transformed mostly into water, carbon dioxide, and other gases.

Nondecomposable wastes such as rock, metal, glass, plaster, and plastic remain buried or stored in the soil. Care must be taken to make sure that liquid and gaseous contaminants derived from wastes decomposing in soils do not contaminate groundwater, lakes, and streams or come to the surface in unacceptable amounts.

Wastewaters and Biosolids

In urban areas, residential and industrial wastewaters are discharged into sewers and are treated at a wastewater treatment plant. The treated wastewater or effluent is typically discharged to a river or lake (or the effluent can be applied to farmland). One of the by-products generated at wastewater treatment plants is sewage sludge, now called biosolids. Biosolids can be dewatered and handled as a solid (10% or more solids) or as a liquid typically over 90% water. Biosolids contain plant nutrients and the organic remains of treated wastewater. Biosolids are typically spread on agricultural land as a source of plant nutrients and carbon for soil organic matter. Biosolids application is not without environmental or health risk; weed seeds, human pathogens, odors, and industrial contaminants may be present in improperly treated biosolids.

Some biosolids from industrial wastewaters contain trace amounts of metals such as cadmium, chromium, lead, and zinc that may contaminate soil and limit the use of biosolids on agricultural land. Dried or composted biosolids (as low as 20% water content) are sold as a fertilizer. It can be a good soil amendment in the same manner as animal manure.

In rural areas, septic tanks and soil absorption systems or drainfields are used to treat residential wastewaters (Fig. 13.6). Wastewater is discharged from a house into a septic tank buried beneath the soil surface, where solids are degraded and the wastewater undergoes primary treatment. The wastewater is then discharged to a gravel bed and the soil beneath the bed is used for final treatment and removal of contaminants. The texture and structure of soil

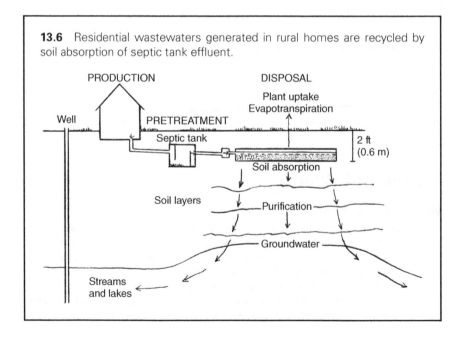

13.6 Residential wastewaters generated in rural homes are recycled by soil absorption of septic tank effluent.

beneath the drainfield and the rate of water discharged determine the degree of treatment. While most pathogens in the wastewater are removed, there is a risk of nitrogen leaching into the groundwater. When soils contain too much clay or gravel or the depth to bedrock or groundwater is too shallow, an artificial soil mound of sandy or local soil material is constructed above the original soil surface. These types of systems remove some pathogens and nitrogen compounds from the wastewater.

Industrial wastewaters vary in composition depending on the process. Some waste waters, such as those generated at canneries, are often spray-irrigated on agricultural land after primary treatment (Fig. 13.7). Many other industrial wastewaters, if not discharged to wastewater treatment plants, are disposed of in stabilization ponds or absorption ponds. The by-products of industrial wastewater treatment such as biosolids or pond bottom sludge may be land-applied if they contain adequate plant nutrients and a minimal amount of industrial contaminants. Stabilization ponds contain impermeable liners and rely on treatment within the pond and evaporation. Wastewater in absorption ponds is treated as it infiltrates through the soil beneath the ponds. Some industries generate hazardous wastes that must be treated and disposed of at special handling facilities. Researchers are currently developing alternatives to these methods of wastewater treatment, including discharge to constructed wetlands.

Solid Wastes

Solid wastes are typically buried in an engineered landfill that is made up of several modules that are filled in sequence over a period of several years. A compacted clay and/or a synthetic liner at the bottom of each module forms a seal to prevent infiltration of liquids that might contaminate the soil and groundwater beneath the landfill.

Precipitation that comes in contact with the solid waste and the liquid generated by the decomposition of solid waste is called **leachate.** Leachate is collected in pipes leading to a containment tank and then treated at a wastewater treatment plant. Solids placed in a cell (small area of the landfill) are compacted and covered with soil daily. The soil on the waste serves several purposes: traps odors, keeps paper and plastic from blowing away, traps contaminants in leachate. When the module is full, it is covered with a clay and/or a plastic cap to minimize the infiltration of precipitation. Several feet of rooting soil is placed as part of the final landfill cover to establish vegetation. Vents are installed in the landfill to allow gas to escape from slowly decomposing waste (Fig. 13.8). Landfill gases that typically contain 65% methane and 35% carbon dioxide are either burned on-site or converted to useable electricity.

Soil and groundwater quality beneath and surrounding a landfill are intensively monitored to prevent environmental contamination. The contaminants of most concern are trace metals, nitrogen compounds, and organic solvents. Due to the large tracts of land needed and the potential for contamination, many municipalities are using alternatives to landfills, such as incineration, composting, and recycling.

Disturbed or Contaminated Lands

Disturbance of land, either by natural processes or human activity, and the contamination of land are often of concern to engineers. Disturbed and contaminated lands result in a situation where soil quality may be impacted, the growth of plants may be severely limited, or plants may not grow at all. Reclamation and remediation of a disturbed area is achieved by restoring it to a productive state. This may include artificial land forming

13.7 Irrigation of farmland is being tried on a limited basis as a means of disposal for wastewater generated by some small industries such as canneries.

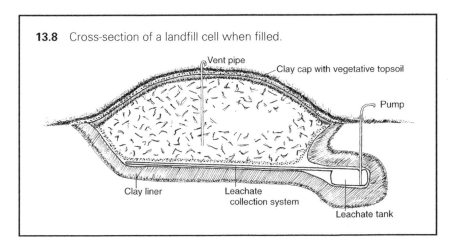

13.8 Cross-section of a landfill cell when filled.

Vent pipe

Clay cap with vegetative topsoil

Pump

Clay liner

Leachate collection system

Leachate tank

and the application of soil amendments and/or fertilizer to be followed by reestablishment of vegetative cover or enhancement of the soil environment to promote degradation of contaminants.

Naturally Disturbed Land

Soil may be disturbed or made unproductive by natural events including landslides; floods that deposit sediment on lowlands; sand dune and dust invasions; blowdown of trees with consequent exposure of soil; and burial of land under fresh lava flows and volcanic ash falls.

Artificially Disturbed Land

Most artificial disturbance of land is caused by human activities, including (1) mechanical strip mining for coal, oil shale, or metallic ore; (2) hydraulic mining of soil material for gold or phosphorus-bearing minerals used in fertilizer; (3) concentration of liquid and solid wastes (including mine tailings) on limited acreage; (4) contamination of soil areas with oil brine around oil wells and with toxic materials near chemical plants such as smelters and oil refineries; (5) contamination of soil beneath and surrounding leaking underground storage tanks; (6) sterilization with residues of agricultural chemicals in low spots in farmland; (7) quarrying; (8) construction (using cut-and-fill operations) in landscapes for development of residential and commercial buildings as well as roads and other facilities in urban areas; (9) operation of vehicles on fragile soils in deserts and tundra areas; (10) overgrazing of rangelands; and (11) overcultivation of croplands.

Many disturbed lands are left in the form of deep holes, mudholes, drifted sand, or mountains of overburden that are steeply sloping and where runoff is rapid. If this is the case, land forming is the first step in reclamation.

Disturbed soils are not necessarily lost to agriculture. At one site in Wisconsin, a sandy clay loam subsoil was needed by a foundry company for making molds in which to pour and cool molten metal. They first removed the topsoil, and then removed the needed subsoil. The fertile loam topsoil was returned to the site, and it now produces high-quality crops just as it did before the excavation (Fig. 13.9).

13.9 Cross-section of parts of two fields. To the left of the post, the soil is undisturbed. To the right, the topsoil was removed and saved, and then the desired subsoil was removed. The original topsoil was replaced, thereby permitting crop production.

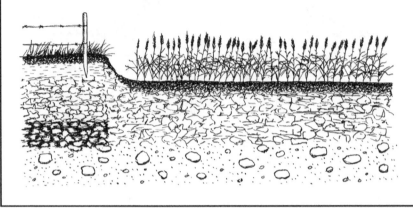

Reclamation and Remediation Procedures

Confinement of Objectionable Substances. Containment of waste-water and biosolids at wastewater treatment plants is necessary to prevent illegal and ecologically damaging spills from flowing into adjacent waters. Fertile agricultural soil itself becomes a polluting waste material if allowed to wash into streams and lakes. In this sense, the reason for preventing soil erosion is to confine the soil by keeping it in place. Brine at oil wells is now commonly pumped into deep layers in the ground, where it can be confined, instead of being poured on the land.

Agricultural chemicals, especially pesticides, should be confined above-ground, where they may be safely used or disposed, to avoid contamination of the groundwater and the resulting pollution of drinking water.

As with biosolids, fly ash from coal-burning electric power plants, if free of heavy metal contaminants, may be spread on farmland as a fertilizer or soil conditioner. If heavy metal contamination is present, those substances must first be removed or the materials must be disposed of in an approved hazardous waste handling facility or landfill.

Removal of Harmful Substances from Soil. Excess salt in soil can be leached downward if there is an adequate supply of freshwater and proper drainage underground or through conduits and ditches to dispose of the salty effluent. It is important that the soils be sufficiently permeable to allow the movement of water through them.

Where chunks of iron sulfide (pyrite and marcasite minerals) in mine tailings yield acid effluent, removal and proper disposal of the sulfide and the neutralization of acid with lime may be implemented.

When the soil beneath underground storage tanks has been contaminated with petroleum products such as gasoline, the soil surrounding the tank may be remediated by using several techniques. The soil is most often excavated and hauled away from the site and spread on agricultural land or placed in a landfill. Many of the contaminants evaporate or are degraded by soil microorganisms. At some sites, the soil is treated in place by enhancing the soil environment for degradation of contaminants (bioremediation). This technique involves injecting air into the soil, which drives some of the contaminants to the surface where they evaporate, while providing an environment more suitable for certain contaminant-degrading microorganisms. Researchers are currently trying to identify soil microorganisms that effectively degrade contaminants and could be introduced into an area where the soil is contaminated. If the groundwater beneath these sites is contaminated, it must also be remediated.

Land Forming. Shaping the land by construction of grassed diversion terraces and waterways can spread runoff and conduct water safely from sloping farmland. Where runoff is caught in holding ponds, the water can be used for irrigation or can simply be allowed to percolate to the water table. Properly maintained terraces can successfully subdivide long slopes into a sequence of short ones, thereby reducing both runoff and soil erosion.

Strip-mined land may be smoothed to slopes that are no steeper than the original ones. Correct stockpiling of topsoil and subsoil during the initial phase of mining makes it possible to restore the topsoil cover.

Some of the sediment that is dredged each year from streams, canals, and catch basins for the benefit of navigation and aquatic life may be useful on agricultural fields. Careful attention must be given to texture, organic matter content, pH, and other physical and chemical characteristics of the dredged materials (spoils).

Some abandoned quarries are filled with soil in such a way as to make them useful for cropping or other purposes. In landscapes with a fairly high water table, standing water is used as a cover for disturbed land. Roadside excavations may become recreational lakes and ponds or sources of irrigation water.

Interim Protection of the Soil Surface. In disturbed land, the soil surface normally needs protection. At some construction sites where disturbance of soil continues for months or even years, mulches of straw or sheets of special fabrics have been used to cover the bare soil until final cover by buildings, pavements, and lawns has been completed.

Establishment of Vegetative Cover. To reestablish vegetation, it is just as desirable to have a good seedbed prepared at the disturbed soil site as it is in a field to be planted to crops. In many cases, this may be difficult because of the nature of the soil material. In most cases, soil amendments and fertilizers may need to be applied. Disturbed soils may be high in some essential elements, but nitrogen is usually quite low.

Reestablishment of native species of vegetative cover by seeding and irrigation has been successful on land that was strip-mined for coal in the Four Corners area of the southwestern United States. In Ohio and Illinois, agricultural crops are being grown today on some prosperous farms that were inactivated for several years by strip-mining operations that ended with reconstruction of the landscape and its soils.

Researchers are currently trying to develop metal-scavenging plants called hyperaccumulators that remove metals from contaminated soil, especially in mined areas, and store them in their leaves and stems. The leaves and stems are harvested and the metals recycled.

Shelterbelts of trees and shrubs illustrate discontinuous establishment of vegetation for the protection of adjacent cropland from wind erosion. The trees reduce the wind speed, thereby reducing the amount of soil detachment and often creating a microclimate that reduces evaporation and increases yields.

Roadside pits from which construction materials have been removed may be stabilized by cut-and-fill operations and subsequent revegetation with trees and/or herbaceous cover. To avoid the need for reclamation, every effort should be made to avoid contamination of the soil and to maintain it in acceptable form. To do this requires knowledge and dedication by those who use the land.

Economic considerations have sometimes led to misuse of the land; thus, government regulations have been necessary to protect the rights of citizens and to ensure productive and beautiful land for future generations.

GLOSSARY

AASHTO—American Association of State Highway and Transportation Officials. The AASHTO system of soil classification is used by engineers for the design and construction of transportation lines (roads, rail lines, and airport runways).

Accelerated erosion—Soil erosion increased by human activity beyond the normal or geological rate.

Acid—A substance with hydrogen ions available for chemical activity.

Acid rock—A rock such as granite that contains considerable amounts of silica and relatively little calcium, magnesium, and iron.

Acid soil—Soil with a pH value of less than 7.0, which is neutral.

Actinomycetes—Threadlike bacteria. Some fix atmospheric nitrogen symbiotically with nonlegume plants.

Aeration, soil—The process by which air in the soil is replaced by air in the atmosphere.

Aggregate—See *Ped.*

A horizon—The natural surface layer of mineral soil that is often referred to as topsoil.

Albic horizon—A light-colored horizon just below the surface from which clays and humus have been leached.

Alfisol—An order of fertile deciduous forest soils that has an accumulation of clay in the B horizon.

Alkali soil—A soil containing sufficient sodium to interfere with the growth of crops (same as sodic soils). The pH is normally 8.3 or higher.

Alkaline soil—A soil with a pH value of more than 7.0, which is neutral.

Alluvial fan—Deposit from a stream as it enters a plain or larger stream.

Alluvial soil—A soil formed from alluvium.

Alluvium—Deposits made by rivers and streams.

Aluminosilicate minerals—Minerals composed largely of oxygen combined with silicon but with some of the silicon replaced by aluminum.

Amendment—Any substance added to soil that alters soil properties, such as gypsum, lime, fertilizer, and sawdust.

Ammonification—The release of ammonia by the microbial decomposition of protein.

Amphibole—A group of dark, basic minerals commonly found in igneous and metamor-phic rocks.

Andisol—The soil order representing soils that developed from volcanic deposits.

Anion—A negatively charged ion.

Anion exchange capacity—The sum total of exchangeable anions that the soil can adsorb. Expressed as centimoles of charge per kilogram ($cmol_c/kg$) of soil material.

Anthropic horizon—A dark surface horizon enriched with phosphorus due to human activity.

Soil Science Simplified, Sixth Edition. Neal S. Eash, Thomas J. Sauer, Deb O'Dell, and Evah Odoi.
© 2016 John Wiley & Sons, Inc. Published 2016 by John Wiley & Sons, Inc.

Apatite—A calcium phosphate mineral used as a source of phosphate for fertilizer.

Aquifer—An underground layer of permeable material that stores and can supply water.

Argillic horizon—B horizon of soil that contains more illuviated clay than the overlying A or E horizon. Some clay coatings are present on surfaces of blocky peds, having moved down from above.

Aridisol—A soil that developed in an aridic environment.

Atmosphere—The layer of gas surrounding the earth: nearly 80% nitrogen, about 20% oxygen, and 0.03% carbon dioxide.

Atterberg limits—Liquid and plastic limits as measured by standard test procedures to determine a soil's suitability as a building material.

Available plant nutrient—See *Nutrients.*

Available soil water—See *Soil water.*

Azotobacter—Free-living bacteria that convert atmospheric nitrogen into organic nitrogen in the soil.

Basalt—A dark (igneous) extrusive rock that is high in iron, magnesium, and calcium.

Bases—Common parlance for ions of calcium, magnesium, and sodium in the soil.

Basic—Another word for alkaline, but often with emphasis on the presence of calcium and magnesium.

Basic rock—A rock with a high content of calcium, magnesium, and possibly iron and a relatively low content of silica.

Bedding—Preparing a series of parallel ridges (beds) usually no wider than that of two crop rows, separated by shallow trenches usually less than the width between crop rows.

Bedrock—The solid rock underlying unconsolidated surface materials.

B horizon—Subsoil horizon found below an A and/or an E horizon.

Biodegradable—Capable of being decomposed into simpler products by living organisms.

Bioremediation—Any of several techniques for optimizing the physical, chemical, and biological conditions in the soil to promote the degradation and/or detoxification of pollutants.

Biosolids—A term for sewage sludge. A by-product of wastewater treatment that contains solids having appreciable amounts of organic material and nutrients but may also contain heavy metals and other potential contaminants.

Bog soil—A peat or muck such as a Histosol.

Boreal—Northern climate that is cool in summer and cold in winter; permafrost is common.

Buffering capacity—Capacity of a soil to resist change, such as a change in the pH.

Bulk density—The mass of a dry soil sample per unit bulk volume (voids and all) as compared with the mass of an equal volume of water.

Calcareous soil—A soil containing enough free calcium (usually also magnesium) carbonate to show effervescence with acid.

Calcic horizon—A subsurface horizon enriched with calcium and other secondary carbonates.

Caliche—A layer near the surface, more or less cemented by secondary carbonates of calcium or magnesium precipitated from soil solution.

Cambic horizon—A weakly developed diagnostic subsurface horizon altered by physical alterations, chemical transformation, or a combination of these.

Capillary action—The action by which the surface of a liquid is elevated or depressed depending on the relative attraction of the molecules of the liquid for each other or a solid with which it is in contact.

Capillary water—Water held in the very small pores of the soil.

Carbon-nitrogen ratio (C:N)—The ratio of the weight of organic carbon to the weight of total nitrogen (mineral plus organic forms) in soil or organic material.

Carbon sequestration—The tying up of atmospheric CO2 in plant tissue.

Carboxyl group—A grouping of carbon, hydrogen, and oxygen (COOH) that is present in organic acids in humus and other materials.

Catena—A sequence of soils from the top of a hill to the footslope.

Cation—A positively charged ion.

Cation exchange—The exchange between a cation in solution with one on the surface of a soil particle.

Cation exchange capacity—The total of exchangeable cations that a soil can adsorb, commonly reported in centimoles of charge per kilogram ($cmol_c/kg$) of soil material.

Channelization—The deepening and/or straightening of natural drainage channels.

Chernozem—Black, organic matter-rich soils common in the subhumid steppes of North America and Asia that are commonly classified as Mollisols.

Chisel—A tillage tool that uses narrow shanks to break up the soil along planes of weakness commonly to a depth of 15 cm. If used deeper than 15 cm, it is usually called subsoiling.

Chlorosis—Lack of chlorophyll in a plant that results in a light green to yellow color of the plant tissue.

C horizon—The soil horizon that is undeveloped the parent material.

Clay—Mineral material composed of particles less than 0.002 mm in diameter.

Coarse earth—The part of mineral soil that is too coarse to pass through a 2-mm sieve.

Colloidal particles—Clay and organic particles that are so small they tend to remain suspended in standing water.

Colluvial deposit—Soil or rock material gathered at the foot of aslope, primarily through the force of gravity.

Compost—Organic residues sometimes mixed with soil that have been piled, moistened, and allowed to undergo biological decomposition.

Conduction—Heat transfer due to the movement of kinetic energy between adjacent atoms in a substance brought about by a temperature gradient.

Convection—Heat transfer through the movement of a fluid (air or water).

Crust of the earth—The outer 12-mile- (19-km-) thick layer of the lithosphere.

Crust—A somewhat dense, hard, or brittle soil layer at the land surface.

Crystalline rocks—Igneous and metamorphic rocks.

Cyanobacteria—Free-living bacteria that convert atmospheric nitrogen into organic nitrogen in crusts that were once called blue-green algae.

Dendritic pattern—A treelike pattern that may be observed at erosion sites.

Denitrification—The microbial conversion of nitrate to the gaseous N form with subsequent release to the atmosphere.

Density—The mass (commonly expressed as weight) per unit volume of a substance.

Diagnostic soil horizons—The soil horizons used for soil classification.

Divalent cation—A cation having two positive charges, such as calcium (Ca^2+).

Dolomite—Calcium-magnesium carbonate commonly called limestone.

Drainage—Water movement through soil. Natural drainage occurs when water drains out of the root zone to deeper layers or to groundwater. Surface channels or subsurface drains (tiles) can be used to artificially drain soils.

Drift—See *Glacial drift.*

Dryland farming—The practice of crop production in low-rainfall areas without irrigation.

Duripan—A diagnostic subsurface horizon cemented mainly with silica.

Eluviation—The removal of soil material in suspension (or solution) from a layer or layers of a soil.

Entisol—A very weakly developed soil; an epipedon underlain by parent material.

Eolian (aeolian) deposit—Materials deposited by wind. See also *Loess*.

Epipedon—A diagnostic surface soil horizon.

Erosion—The process by which soil is washed, blown, or otherwise moved by natural agents from one place on the landscape to another.

Escarpment—A cliff or very steep slope that is fairly continuous and found at the edge of an area with more gentle slopes; often the result of differential erosion rates of previously overlying strata.

Eukaryotes—Organisms whose cells have nuclei may be plant or animal.

Evaporation—The change of the state of water from a liquid to a gas, as happens at the surface of bare soil.

Evapotranspiration—The transfer of water vapor to the air by a combination of evaporation and transpiration.

Exchange complex—Surface of clay and humus having primarily negatively charged sites in most soils.

Exchangeable cations—Cations such as those of calcium, magnesium, and potassium that are held loosely enough on surfaces of colloidal soil particles that they can exchange places with cations in the soil solution nearby.

Fauna—Animals present at a site or in a region.

Feldspar—The most common primary mineral in the earth's crust.

Fertility, soil—The status of a soil with respect to its ability to supply the nutrients essential to plant growth.

Fertilizer—Any organic or inorganic material of natural or synthetic origin that is added to a soil to supply one or more elements essential for the growth of plants.

Field capacity—The percentage of water by weight that is held in the soil by capillary action after free drainage by the force of gravity has practically ceased.

Fine earth—That portion of a soil that is finer than 2 mm in diameter, including mineral sand, silt, and clay.

Flora—The plants present at a site or in a region.

Fragipan—Very dense subsoil layers that are brittle when dry, but not when wet, primarily cemented by silica.

Furrow diking—Creating a dike and a small depression in furrows (normally 6-12 ft [1.8-3.6 m] apart) to hold water and reduce runoff.

Gabbro—A dark-colored crystalline igneous rock.

Gelisols—An order of soils characterized by having permafrost.

Genesis of soil—Soil formation.

Geomorphology—The study of landforms on the surface of the earth.

Gilgai—See *Vertisol*.

Glacial drift—Deposits made by glaciers and their meltwaters, including till and outwash.

Glacial till—Unsorted debris left by a glacier.

Glacier—A large body of slowly moving ice.

Global positioning system (GPS)—A global navigation system that uses satellites to triangulate positions on earth.

Gradient—A measure of soil in feet (meters) of rise or fall per 100 ft (31 m) of horizontal distance.

Granite—A light-colored, crystalline igneous rock containing considerable quartz (about 25%).

Green manure—A growing crop that is plowed under and mixed with the soil to enrich it with organic matter; also used to promote plant uptake of mobile nutrients between seasons.

Groundwater—Water beneath the earth's surface in saturated soil or porous rock strata.

Gully erosion—Removal by water of sufficient soil to form channels large enough to prevent machinery from crossing

Hardpan—A soil layer that acts as a barrier to the movement of water and plant roots.

Heat of fusion—The amount of heat necessary to change a substance from its solid to its liquid phase.

Heat of vaporization—The amount of heat necessary to change a substance from its liquid to its vapor phase.

Hectare—An area of land equal to 10,000 sq m or 2.47 acres.

Histic epipedon—An organic surface horizon that is characterized by saturation for prolonged periods unless artificially drained.

Histosol—A soil order of peats and mucks that are thicker than 18 in. (46 cm).

Horizon—A soil layer that formed parallel to the land surface during the natural development of the soil body.

Humus—The dark, rather stable part of soil organic matter that remains after the major portion of animal and plant residues have decomposed and disappeared in the form of water and gases such as carbon dioxide.

Hydration—A mineral weathering process wherein water molecules combine with soil constituents.

Hydrologic cycle—A cyclic pathway that describes water movement on earth. Includes such processes as evaporation, sublimation, precipitation, runoff, percolation, and groundwater flow.

Hydrolysis—A chemical decomposition reaction in which a chemical bond is split and the elements contained in water are added.

Hydrous mica—A common layer silicate clay that is intermediate in responsiveness to wetting and drying between smectite and kaolinite.

Hyphae—Individual threads of mycelia.

Igneous rock—A rock that formed by the cooling and solidification of liquid parts of the lithosphere.

Illite—A hydrous mica type of silicate clay.

Illuviation—The process of deposition of soil material removed from one horizon to another in the soil; usually from an upper to a lower horizon in the soil profile.

Immobilization—The conversion of inorganic ions from the soil into organic molecules in living tissue. Changes available nutrients into unavailable forms.

Inceptisol—An order of moderately developed soils. It consists of an epipedon underlain by a cambic horizon and parent material.

Infiltration —The downward entry of water into soil.

Infiltration rate—The rate at which water can enter the soil under specified conditions, including the presence of an excess of water.

Interveinal—Between the veins.

Ion—An electrically charged atom, with a surplus or deficiency of electrons.

Ironstone—See *Plinthite*.

Kaolinite—A 1:1 layer silicate clay with a low degree of responsiveness to wetting and drying and low nutrient-holding capacity.

Landform—A natural feature of the earth's surface, such as a hill or a plain.

Latent heat—The amount of energy used to change the phase of a substance.

Laterite—A surface formation rich in aluminum and iron that developed through the weathering of the underlying parent rock. It has been used extensively in building throughout the tropics.

Lattice—A three-dimensional grid of lines connecting points that represent the centers of atoms or ions in a mineral clay.

Layer silicate clay—See *Silicate clay*.

Leaching—The removal of materials in solution by downward movement of water through soil.

Lime—Ground limestone, either calcite or dolomite.

Limestone—A rockrich in calcium carbonate derived from shells of sea organisms and the mineral calcite, including dolomite, which is rich in calcium and magnesium carbonate.

Limy soil—A soil containing appreciable amounts of carbonate minerals such as calcium or magnesium carbonate or a combination.

Lithosphere—The solid, rigid rock portion of the earth.

Loess—Deposit of windblown soil particles, largely silt size.

Macronutrient—A chemical element that is essential for plant growth and is used in relatively large quantities (usually >50 ppm in plants).

Magma—Liquid rock found at great depths in the earth's crust or exposed during volcanic eruptions.

Manure—Excreta of animals, with or without bedding material, normally added to soil to improve it with respect to crop production. See also *Green manure*.

Mass wasting—The movement downslope under the pull of gravity of large masses of soil and/or rock.

Matric force—The force of attraction between water and soil particles that holds capillary water in the soil.

Matrix—Something within which something else originates or develops.

Melanic horizon—A dark, organic matter-enriched surface horizon in some soils formed in volcanic deposits.

Metamorphic rock—Rock formed by recrystallization of igneous or sedimentary rock under great pressure and heat whereby the minerals contained in the rock become reoriented.

Micronutrient—A chemical element necessary for plant growth and used in relatively small amounts (usually <50 ppm in the plant).

Microrelief—Slight irregularities of a land surface.

Mineral—A natural inorganic compound with known physical properties and chemical formula, a mixture of which forms a rock.

Mineralization—The microbiological decomposition of organic matter into inorganic products.

Mites—Arthropods that are abundant in soil.

Mollic epipedon—Dark, thick, organic matter-rich, fertile topsoil commonly found in the subhumid grasslands of Asia and North America.

Monovalent cation—A cation having a single positive charge, such as sodium (Na+).

Montmorillonite—A very reactive 2:1 layer silicate clay. A member of the smectite group of clays.

Moraine—A blanket or ridge of unsorted debris left by a glacier.

Mottling—Spotted areas of color in a soil, usually associated with periodic wet conditions.

Muck soil—An organic soil that is more than half decomposed few visible organic fibers constitute a minor portion of the mass.

Mulch—A layer of material spread over the soil surface to protect it and plant roots from erosion, crusting, freezing, and drying.

Munsell colors—A system for uniformly describing the color of soil.

Mycelium—A threadlike mass of fungal hyphae.

Mycorrhizae—Fungi that live symbiotically with the roots of higher plants.

Myxomycetes—Slime molds; morphologically between bacteria and fungi.

Necrosis—Death or decay of plant or animal tissue.

Net radiation—The difference between incoming and outgoing short-wave and long-wave radiation.

Nitrification—Biological oxidation of ammonium to nitrite and nitrate.

Nitrobacter—Bacteria that perform the second and final step of nitrification in the nitrogen cycle.

Nitrogen cycle—The biochemical changes that take place in repetitive sequence as organisms take up and release nitrogen.

Nitrogen fixation—Biological conversion of molecular dinitrogen (N_2) to organic combinations utilizable in biological processes.

Nitrosomonas—Bacteria that perform the first step of nitrification in the nitrogen cycle.

Nutrients—Substances essential for the growth of plants, such as nitrogen, phosphorus, and potassium. An available nutrient is one that can be readily absorbed by the plant.

Ochric epipedon—Surface soil that fails to meet the criteria for any other epipedon.

O horizon—A thin, surficial organic layer on soil, such as leaf litter.

Organic gardening—Gardening without the use of commercial fertilizers and pesticides. Emphasis is on the use of composted materials, green manures, and judicious arrangement of plants and associated beneficial organisms.

Organic soil—A soil that contains a high percentage of organic carbon.

Ortstein—A natural soil pan formed in spodosols (Podzol soils) by cementation of the subsoil by organic matter and iron oxide.

Osmosis—The diffusion of water through a differentially permeable membrane, such as a root hair, from an area of high water concentration to an area of lower water concentration (or low salt concentration to high salt concentration) if pressure and temperature are equal on each side of the membrane.

Outwash—Deposit made by flowing meltwaters from glaciers.

Oxic soil horizon—A relatively infertile subsurface diagnostic horizon common in tropical areas.

Oxidation—A mineral weathering process wherein oxygen ions combine with multivalent elements such as iron, loss of electrons from an atom, molecule, anion, or cation.

Oxide clay—Fine particles composed of oxides of iron and aluminum, commonly in noncrystalline or amorphous forms.

Oxisol—A soil order common in tropical areas that is relatively infertile.

Parent material—Consolidated or unconsolidated material from which the soil develops through pedogenesis.

Particle size distribution—Proportion of clay, silt, and sand particles in the fine earth of a soil.

Peat—A soil with relatively undecomposed fibrous organic material.

Ped—A unit of soil structure in a soil horizon may be blocky, platy, or granular. A soil aggregate is the same as a soil ped.

Pedon—A volume of soil as it occurs in nature that is large enough to show the variations in the horizons in the third dimension usually an area of 1-10 sq m.

Percolation—The downward movement of water in soil.

Petrocalcic horizon—A highly developed and cemented calcic horizon, often called caliche.

pH—The degree of acidity or alkalinity. The hydrogen potential expressed by a set of negative logarithmic values whereby numbers less than 7.0 signify acidity and numbers greater than 7.0 signify alkalinity. pH 7.0 is neutral, that is, neither acid nor alkaline.

Plinthite—A nonindurated mixture of iron and aluminum oxides, commonly with some quartz and kaolinite clay, that has the capacity to harden irreversibly into ironstone upon repeated wetting and drying.

Podzol—A great soil group under the FAO system, now replaced by the soil order Spodosol in the U.S. system of Soil Taxonomy.

Pollution—The act of polluting, that is, to contaminate, make unclean or impure.

Polypedon (soil body)—Contiguous, similar soil pedons constituting a unit of land to the depth of soil development.

Porosity—The volume percentage of the total bulk not occupied by solid particles.

Prairie soil—Mineral soils with organic matter-rich mollic epipedons resulting from the annual necrosis of grass roots, commonly classified as Mollisols.

Precipitation—Rainfall and snowfall plus minor amounts of dewfall and fog drip.

Primary mineral—A mineral formed naturally by crystallization from molten rock. Feldspar is the most common primary mineral in the earth's crust.

Productivity—The capacity of a soil to produce plant material. This is expressed in yield per unit of area per unit of time.

Profile —A two-dimensional, vertical cross-section of a soil through all horizons.

Prokaryotes—Simple organisms whose cells lack nuclei not clearly plants or animals.

Quartz—A resistant primary mineral consisting of silicon dioxide common in most rocks.

Quartzite—Metamorphozed sandstone.

Radiation—Heat transfer by electromagnetic waves.

Reduction—An anerobic mineral weathering process that results in the gain of electrons to the atom, ion, or molecule.

Regolith—Unconsolidated material above solid bedrock.

Relief—The elevation, or differences in elevation considered collectively, of a land surface on a broad scale.

Rhizobia—Bacteria that convert atmospheric nitrogen into organic nitrogen in root nodules of legumes.

Rhizosphere—A zone of the soil where plant roots are in abundance and soil microbes are especially active.

Rhyolite—A fine-grained extrusive equivalent of granite.

Rill erosion—Water removal of soil in channels small enough to be filled in by normal tillage.

Runoff—Water from precipitation or snowmelt that flows over the soil to surface water bodies.

Salic horizon—A subsurface diagnostic horizon enriched with soluble salts.

Saltation—The leaping, jumping, or bouncing of soil particles along the surface of the soil during wind erosion.

Sandstone—A sedimentary rock, usually of quartz, bound together by a cementing material such as silica or iron oxide.

Saprolite—A mass of weathered rock.

Saturation—The condition of a soil when all pores are filled with water.

Secondary mineral—A mineral formed by weathering from a primary mineral. For example, kaolinite is a secondary mineral formed from feldspar. All carbonates such as calcite and dolomite are secondary minerals.

Sedimentary rock—A rock composed of sediment, that is, deposits made by water, wind, ice, and gravity.

Sensible heat—The amount of energy used to warm the air above the soil surface.

Shale—A sedimentary rock made of clay, silt, and very fine sand.

Sheet erosion—Water removal of a thin layer of soil from the soil surface. Results in the highest erosion losses on an area basis and difficult to visually detect.

Silicate clay—Clay composed of minerals made up largely of crystalline layers of silica and alumina.

Slickenside—Polished surfaces caused by blocks of soil sliding past each other during the formation of Vertisols.

Smectite—A group of very reactive silicate clays of which montmorillonite is aprominent member.

Sodium adsorption ratio (SAR)—The ratio of soluble sodium to soluble calcium and magnesium. SAR is used to assess the quality of irrigation water and the risk of soil dispersion.

Soil—*Definition 1:* The unconsolidated mineral or organic material on the immediate surface of the earth that serves as a natural medium for the growth of land plants. —*Definition 2:* The unconsolidated mineral or organic material on the surface of the earth that has been subjected to and shows the effects of genetic and environmental factors of climate (including water and temperature effects) and macro- and microorganisms, conditioned by relief, acting on parent material over a period of time.

Soil absorption system—A system of pipes buried in the soil through which effluent from septic tanks infiltrates into the soil.

Soil body—A component or unit of the natural terrain (landscape).

Soil classification—Grouping of soils based upon common characteristics such as diagnostic surface and subsurface horizons.

Soil formation (genesis)—The pedogenic processes by which parent material is transformed into a body of soil.

Soil heat storage—The amount of energy stored as heat in the surface soil layers.

Soil landscape—The soil portion of the landscape.

Soil management—The sum total of all tillage operations, cropping practices, fertilizer, lime, and other treatments conducted on or applied to a soil for the production of plants.

Soil moisture regimes—A system for describing the predictable available moisture in the crop root zone of the soil throughout the year.

Soil reaction—Degree of acidity or alkalinity (pH).

Soil sequence—An array of soils (soil bodies) such as from the top of a hill to the footslope or from the youngest soil on the youngest landform surface to the oldest soil in the same region.

Soil solution—The aqueous liquid phase of the soil and its solutes.

Soil structure—The clustering of soil particles into units called peds or aggregates.

Soil survey—The identification, classification, mapping, scientific and practical explanation, evaluation, and interpretation of the soil cover of a terrain.

Soil temperature regimes—A system for describing the predictable soil temperature throughout the year. It has implications for crop production.

Soil texture—The relative proportions of the three soil separates (sand, silt, and clay).

Soil water—Water contained in soil. Available soil water is that which can be absorbed by the plant.

Solum—The A and B horizons together or the single one of these that overlies the C horizon at a site.

Spodic horizon—B horizon in which sand grains are coated with aluminum oxide, humus, and usually iron oxide. It is best developed under acidic tree litter as is common in coniferous forests.

Spodosol—A soil order common in humid areas under acidic forest horizons in which the spodic horizon is prominent.

Stomates—Minute openings in a leaf that permit gaseous interchange.

Stratification—Layering of water-laid deposits such as sand and gravel.

Stubble mulch—The stubble of crops or crop residues left essentially in place on the land as a surface cover before and during the preparation of the seedbed and possibly during the growing of a succeeding crop.

Sublimation—The change of state for water from a solid to a vapor. The transformation of snow or ice directly to water vapor.

Substratum—A layer beneath the surface soil.

Summer fallow—Water-conserving shallow tillage that prevents vegetative growth during alternate cropping seasons in arid and semi-arid regions.

Surface creep—The rolling by wind of coarse sand particles over the soil surface.

Suspension—The movement by wind of small soil particles (silt and clay) caught up by air currents.

Symbiotic nitrogen fixation—The close association of certain bacteria and legumes to convert atmospheric nitrogen (N_2) into nitrogen forms usable in biological processes.

Tectonic activity—Disruption of the earth's crust, resulting in earthquakes, volcanoes, faults, and related events.

Terrace—A benchlike landform on valley bottoms ("high bottoms").

Thermal conductivity—Property of a substance that indicates its ability to conduct heat.

Tillage—Mechanical manipulation of soil for any purpose. In agriculture it is usually restricted to the modification of soil conditions for crop production.

Tilth—The physical condition of soil as related to its ease of tillage and fitness as a seedbed.

Topography—The relief of the land, that is, levelness or hilliness.

Transpiration—The transfer of water to the atmosphere through the stomates of plant leaves.

Ultisol—A soil order of soils in warm region forestlands with an argillic diagnostic subsurface horizon, less base saturation than alfisols.

Umbric epipedon—Dark, deep, surface soil horizon that is more acidic than a mollic epipedon.

Unsaturation—The state of a soil in which most of the smaller pores are filled with water and the larger ones are primarily filled with air.

USC (Unified Soil Classification) system—The USC system of soil classification is used by engineers involved in foundation engineering.

Vermiculite—A 2:1 layer silicate clay derived from hydrous mica.

Vertisol—A soil order high in expanding smectitic clays that invert during repeated wetting and drying cycles causing a wavy surface microtopography (gilgai).

Watershed—An area draining ultimately to a particular body of water.

Water table—The surface of the groundwater.

Weathering—The physical disintegration and chemical decomposition of minerals and rock.

Wilting point—The condition of a soil when its water content is so low that plant roots can no longer obtain adequate water to sustain life.

Windbreak—Vegetation planted to protect downwind crops from desiccation and breakage and to protect soil from wind erosion. It is also used to protect homesteads on the High Plains from high winds.

INDEX

Soil Science Simplified, Sixth Edition. Neal S. Eash, Thomas J. Sauer, Deb O'Dell, and Evah Odoi.
© 2016 John Wiley & Sons, Inc. Published 2016 by John Wiley & Sons, Inc.

Printed and bound by CPI Group (UK) Ltd, Croydon, CR0 4YY

16/04/2023

03210842-0002